Conrad Merk

Excavations at the Kesslerloch Near Thayngen, Switzerland

A cave of the Reindeer period

Conrad Merk

Excavations at the Kesslerloch Near Thayngen, Switzerland
A cave of the Reindeer period

ISBN/EAN: 9783337196899

Printed in Europe, USA, Canada, Australia, Japan

Cover: Foto ©ninafisch / pixelio.de

More available books at **www.hansebooks.com**

EXCAVATIONS

AT THE

KESSLERLOCH NEAR THAYNGEN

SWITZERLAND

A CAVE OF THE REINDEER PERIOD

BY

CONRAD MERK

TRANSLATED BY

JOHN EDWARD LEE, F.S.A., F.G.S.

AUTHOR OF "ISCA SILURUM" ETC.
TRANSLATOR OF KELLER'S "LAKE DWELLINGS"

LONDON
LONGMANS, GREEN, AND CO.
1876

PREFACE

SOME LITTLE TIME AGO my kind friend Dr. Ferdinand Keller, the President of the Society of Antiquaries at Zürich, who frequently keeps me informed of what is going on in the archaeology of his district, sent to me the first part of the Proceedings of his Society for the present year. It contained an account of the Excavation at the Kesslerloch, a cave of the Reindeer period near Schaffhausen. The moment I saw it it struck me that it was of very great interest; and it occurred to me very forcibly that it ought to be translated into English. This idea was communicated to Dr. Keller, and he at once assented, and very kindly allowed me to have lithographic transfers from the Society's plates; so that the drawings now before the reader are the actual original sketches.

Such is the origin of the present little volume. I cannot think that I am wrong in considering the excavation in this cave as one of the most interesting which has been made for years; and though it must be confessed that the following report is by no means perfect—that it is defective in some parts and redundant in others—yet it is a wonderful production for a young author, for such in fact is the excavator of the Kesslerloch.

In certain cases where it appears to me that the author is not quite correct I have ventured on adding notes stating my opinion: and I have also gone so far as to alter in a very small degree the style where it appeared to me to be inappropriate. Otherwise the whole report is left precisely in the words of the author, although to regular antiquaries it may appear that it might have been shortened with advantage.

It would be hardly correct if in a preface to this most interesting account I omitted to state the astonishment which these prehistoric drawings have caused in certain quarters, almost amounting to scepticism as to their genuineness. But, on the other hand, it may perhaps be allowable to state a few facts which somewhat bear upon this point. In the course of the past summer, finding myself in Germany, it seemed very desirable to see with my own eyes the objects which had been found at the Kesslerloch, and also to visit the place itself. When at Zürich I had the good fortune to meet with Mr. Franks, who, it is well known, has paid very great attention to prehistoric matters—in fact he had been attracted to Switzerland by these very discoveries. With great kindness he allowed me to accompany him to Schaffhausen, where a large portion of these relics are deposited in the Museum there, and also to the cave itself. Dr. Von Mandach, the President of the Schaffhausen Natural History Society, accompanied us. The sketch forming the frontispiece was taken during this hasty visit, and was afterwards corrected by a second glance.

It will, I hope, not be a breach of confidence if I mention the facts as to the excavation which came to my knowledge. Thayngen is reached by the railway from Schaffhausen in little more than a quarter of an hour. Mr. Merk, the author of the present report, was for some time the Government 'teacher of practical science' there: and when he had discovered the cave, it was arranged with the Schaffhausen Society, of which the active and intelligent Dr. Von Mandach, the leading physician of the place, is the President, that the Society was to pay the whole expense of the excavation, and to receive one half of the objects found; the other half was to repay Mr. Merk for the trouble he took in superintending the excavation.

It may be well imagined that Dr. Von Mandach and his coadjutors at Schaffhausen looked very narrowly after the products of the excavation, for the cave was only a few minutes distant from them by railway, and the Society had engaged to spend a considerable sum of money. The progress of the excavation is given in the following pages; and it seems to me that if evidence is of any value at all, we have here one of the strongest cases in which antiquities, excavated under proper superintendence, may be considered as genuine.

One half of the collection was sold by the discoverer to the Constance Museum; the other half remains in the Museum at Schaffhausen. This portion was carefully examined by my archaeological friend and myself, and I believe I am correct in saying that neither of us have the slightest doubt as to its being genuine. The mere fact of the drawings being so much better executed than those of Aquitaine (a fact, however, which Mr. Franks doubts) is no real argument against them, as the objection on this score arises merely from a preconceived but improved notion that all cave-dwellers must have been of so low a grade as to be totally incapable of any knowledge of art.

There are, however, two specimens which are not included in the collection at Schaffhausen, and respecting these it is perfectly natural to have grave doubts. They have been drawn by Mr. Merk, and also by Professor Rütimeyer in his late work on the Swiss Fauna, as will be seen in the Appendix. The drawings are therefore reproduced here. The story of the discovery of these two specimens is given in a note appended to the description, and I have not scrupled to tell the whole truth, so far as it can be ascertained, as in a matter of this kind perfect honesty is required. I will merely add, that although individually I rather incline to think that one of them at least is genuine, yet they must stand on their own merits, and on their own merits alone. It may be well to add that these two specimens, shaded as they are by the cloud about their origin, have, by the liberality of my archaeological friend above mentioned, been placed for reference in the collection of prehistoric objects in the Christy Collection, Victoria Street.

JOHN EDWARD LEE.

Villa Syracusa, Torquay :
October 21, 1875.

CAVE OF KESSLERLOCH.

—⸺⊷⧫⧫⧫⊶⸺—

THE first time I entered the Cave of Kesslerloch was in the
summer of 1873, when on a botanical excursion. It was then
completely overgrown with trees and shrubs: in fact in summer
the interior was altogether closed to the chance passer-by. As
I had been stationed at Thayngen hardly a year, the cave was
entirely unknown to me, and probably would have remained so,
but for the fact that the *Althæa officinalis* flourished in the
neighbourhood in great abundance and of unusual size, and this
led me to it. I could not resist securing such specimens, and
thus was in a position to notice the yawning background of the
cavern through the bushes, where they were least dense.
Having pushed with some difficulty through the foliage, I was
not a little astonished to find that behind all this life and
verdure I was surrounded by bare walls of rock. For some
years past I had given much attention to geology and the study
of prehistoric man, and I was strongly impressed with the idea
that this cave, like those of other countries, might have been
the abode of some of the human race in prehistoric times. I
then formed the determination to make excavations. But the
realisation of this intention had to be put off for some time.
The summer and the autumn passed away, and the ground was
already covered with light snow, when a lecture which I had
to give on volcanos reminded me strongly of the resolution I
had made, but which had well nigh faded away from my
memory. I communicated my ideas to my colleague, Mr. Wepf,
who confirmed me in my determination, and offered to join me
in the excavation.

Thus on December 4 we went to the Kesslerloch with two
of our older pupils, shovels and pickaxe in hand. The ground
was already hard frozen, but by chance we dug a trench in a
very instructive place. For some time we obtained none of the

results we hoped for, and it was not till we had dug to a depth of about 39 inches that we found the slightest trace of animal bones; but amongst those we then discovered were some very large horse's teeth. After continuous work for about three hours we returned home with a rich store of bones, still uncertain whether our labour had been bestowed simply on a bone cave or on a human habitation. This uncertainty was naturally not agreeable to us, and we continued our excavations still deeper, and then we discovered some few flint-flakes, and also the first reindeer-horn, which, on a close examination, bore undoubted marks of having been worked. Thus it was proved beyond a doubt that the Kesslerloch had been inhabited in prehistoric times. It may be imagined that our delight at this discovery was not small; and thus, in order, on the one hand, to make it perfectly certain, and, on the other, to keep away all foreign elements from the excavation, we secured from the owner of the Kesslerloch, for a sum of money by no means inconsiderable, the right of carrying on excavations in the cave exactly in the manner we pleased.

The regular systematic excavations began on February 19, 1874, and with little interruption continued till April 11—in other words fully seven weeks; and on the average five men were at work the whole time : all of them without exception worked most diligently. It seems only right that I should here mention the intelligent and untiring labours of Mr. Schenk von Eschenz, of the Canton Thurgau, who in a most disinterested manner gave himself up to the investigation, and has earned our most heartfelt thanks for his valuable services.

The Natural History Society of Schaffhausen, by its courteous and valuable assistance, and more especially by the great trouble taken by the President, Dr. von Mandach, enabled us to push the work on more quickly, and to carry it on in a more systematic manner; and this Society also undertook all the cost of the excavation ; so that we may venture positively to affirm that the cave has been excavated most completely, and in a highly satisfactory manner.

It received the name of Kesslerloch, as I have been informed, from its having been the abode about fifty years ago of a family of tinkers or wandering smiths.[1]

---

[1] The word 'Kes-lerloch,' strictly translated, would be the 'tinker's cave' or 'hole;' but as the name has to be repeatedly used in the course of the following pages it seems better—or at any rate it *sounds* better—to retain the German word 'Kesslerloch.'

It is situated about ten minutes' walk west of Thayngen, rather a large place in the Canton of Schaffhausen; it lies close to the borders of the Grand Duchy of Baden, and is very near a well-made tunnel on the railroad leading from Constance to Schaffhausen. In front of the cave there are two narrow pleasant vales, uniting at right angles, the bottom of which consists of meadow-land, while the sides are formed partly of naked water-worn rocks, but chiefly of slopes from 60 to 80 feet high, covered with light underwood. The side valley, to the north, which here is about 98½ feet long by about 200 feet wide, joins the main valley leading from Thayngen, by Herblingen, to Schaffhausen, where there is a small brook, a narrow footpath, and the line of railroad, and here is to be found the last outpost of the Swiss Jura, about 30 feet high, and very steep towards the side valley. In this steep wall of rock, consisting of white Jura, or oolitic limestone, the Kesslerloch is situated, much about the same level as the flat of the valley.

Description of the Cave.—There are two openings into it; that to the south, 10 feet broad, is about 7 feet above the

Fig. 1.

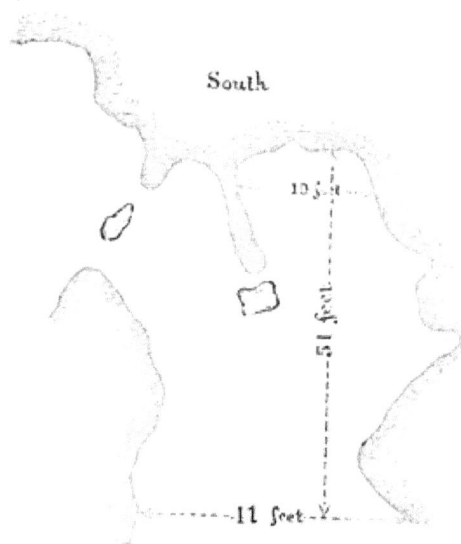

South

10 ½ ft

51 feet

11 feet

level of the valley, and towards the west widens out into a cavity from 3½ to 5 feet high and 23 feet long. A gentle slope, formed by a number of fallen limestone fragments, covered with low brushwood, makes the entrance perfectly easy. The

chief opening, on the east side, in the perpendicular rock is
now visible from a distance; formerly it was hidden by the
brushwood. It is 11 feet wide and $11\frac{1}{2}$ feet high. This
entrance is level, and dry underfoot, so that there is no difficulty in making use of it.

The length of the cave is about 51 feet, but inside both
the breadth and height lessen rapidly, so that at 25 feet from
the opening the breadth is only about 31 feet, and the height
hardly 6 feet. In the middle there is a pillar of weather-worn
limestone about 20 feet in circumference, which, together with
a kind of screen or partition hardly a foot thick and 30 inches
high, divide the back part of the cave into two smaller portions. Between the pillar and the screen there is a roundish
opening. The northern portion of the cave is about 21 feet
long and 19 feet broad, but it is hardly 39 inches high, so that
the excavation here was attended with some difficulties. The
southern part has nearly the same dimensions, but it contracts
towards the southern opening, which can only be entered
when stooping down. The sides and roof are covered with
numerous little cracks or fissures, and have a number of niches
or small projections of different sizes; these, together with the
number of stalactites of various sizes in the hinder or northern
portion of the cave, give to it a decidedly picturesque appearance. The superficies of the ground actually covered over
by the cave is about 2,000 square feet, and the cubical contents
of the whole cave about 10,500 cubic feet, so that the Kesslerloch in superficial extent is about seven times as large, and
in its cubical contents more than four times as great, as a
room 19 feet 8 inches long, 14 feet 9 inches broad, and 8 feet
10 inches high. The great agreeableness of the situation, the
picturesque appearance, the convenient space, and the excellent
supply of light in the cave rendered it a desirable abode even
for men of the nineteenth century, accustomed as they are
to so many conveniences. How much more would this have been
the case thousands of years ago as a place of refuge and a
permanent abode for these Troglodytes!

We will now refer to the succession and character of the
different beds in the cave; and first we may mention as the
uppermost bed a mass of rubbish formed both of small and
large angular stones of white Jura limestone, like the rocks
on either side. There can be no doubt that these fragments
of limestone, not having been rounded or worn by water,
have been derived from the rocky walls of the cave, which

when the percolated water within them froze, became cracked, and then crumbled; and thus gradually in the course of many centuries a bed of rubbish was formed of various degrees of thickness. This bed extends more especially in front, where the rocks were more exposed to the weather, and here its greatest thickness amounted to from 47 to 55 inches, while behind it gradually became thinner, so that in the middle it was about 41 inches thick, and in the further northern chamber only about 3½ inches. In that part which lay towards the southern entrance it was hardly 2½ inches in thickness. This may possibly be accounted for by the decided slope of that part of the cave. In this rubbish, and more especially in front towards the eastern entrance, lay a number of large stones also consisting of white Jura or oolitic limestone. One of these stones was remarkable for its great size, being 73 inches long, 59 inches broad, and 53 inches high, so that the weight may be estimated not far from 5 tons. If it had been the case that a human being of that period had been buried under these stones when they fell we should have had tangible evidence of human occupation; but chance has not so willed it.

It took a long time to clear away all the upper bed of rubbish, as it contained nearly 4,300 cubic feet. But this covering was very useful in preserving the materials lying beneath; for if the whole weight be calculated it amounted to about 170 tons, thus exerting a pressure of about 1¾ cwt. on each square foot, so that the relic-bed lying beneath was very much pressed down, and thus was less exposed to decomposition. In different places some inches under this rubbish-bed there lay two beds of stalagmite: one was about 54 square feet in extent, stretching from the northern side of the northern area, and was from 12 to 18 inches thick; the other formed a band alongside of the south wall of the rock, and was 18 or 19 inches thick. The stalagmite was so hard that it had to be blasted with gunpowder in order to separate it from the underlying relic-bed. Both these beds of stalagmite contained on the underside a great number of bones and a few flints, evidently showing that the stalagmite had begun to form when the cave was inhabited.

As before mentioned, under this rubbish-bed there lay a *black bed* called the relic-bed (*Kulturschicht*), because it contained so many relics of early civilisation, and a mass of bones of different animals long since passed away. These silent evidences of prehistoric times lay scattered one from another in a bed of large and small stones crumbled down from the oolitic limestone above. In front this bed was 15 inches thick, in the middle of the cave about 11 inches, and in the back part only about 4 inches thick. This decided diminution in the thickness of the relic-bed may be accounted for partly because the front portion of the rock would weather to a greater extent and more rapidly, and partly because the heaps of bones diminish towards the back portion; and this again may be accounted for by supposing that the early inhabitants of the cave had their work-room in front, and passed most of their time there, while the hinder portion may have been chosen for their sleeping apartment. The relic-bed, the black colour of which arises from the slow decomposition of animal substances, extends over the whole of the cave, even quite under the pillar before mentioned,[1] so that we cannot imagine that there was any particular place where the rubbish was thrown. What could not be eaten was cast aside, without taking any account of the unpleasant effluvium which would arise from the decay of the meat, which in part still adhered to the bones.

The bones in this bed, which were of a yellowish white colour, were easily preserved, but some few of the bones were so rotten that they crumbled to pieces in the hand. Many of the bones were either partially or entirely infiltrated with iron and manganese forming dendrites. In several places of the relic-bed there were hearths of different sizes, round which in general there were several slabs, which probably were used as seats for those who were sitting round the fire.

On the north side of the cave there were three rather large slabs of what is called 'oolitic marble,' imbedded in fine rich loam. These may probably have been used as raised sleeping places or couches. Who had the distinction of occupying them, whether all the members of the clan, or merely the chief, we must leave undecided. At all events, I cannot agree with the view that these slabs were used as hearths or places of offering, for neither the slabs nor the loam show any trace of the action of fire. Neither can these slabs have been used as tables for working the loam upon, for not a single fragment of pottery was found in the whole of the relic-bed. It is quite

---

[1] Is it not possible that this may be a mistake? My own impression when I saw the cave was that the pillar was a portion of the regular oolitic rock; but my visit was hasty, and I may be in error.

clear that the slab is of the same age as the bones and implements found here, for a narrow strip or band of the relic-bed runs quite underneath it, while round about it it is much thicker. It is a peculiar and interesting fact that the relic-bed in front goes down deeply under the present surface of the valley; this has led me to many considerations as to the age of this bed, to which I hope to refer subsequently. Under the relic-bed, and extending over the whole of the cave, there was another bed *coloured red* with oxide of iron, and which consisted, like the others, of crumbled limestone. This bed also contained a number of bones and implements which were better preserved than the others, as they were constantly surrounded with the water of the soil, and thus were less exposed to the decomposing influences of atmospheric air. Here I may remark that water is very essential for the preservation of bones. Doubtless in prehistoric times every convenient cave was inhabited; but from the want of water the human and animal remains in the course of thousands of years have disappeared from the majority of these caves, so that with the exception of the flint-flakes every indication of the former inhabitants has been lost. The cave lately found in the neighbourhood of Berne is a proof of this.

The thickness of this second relic-bed varied from 14 inches to 2½ inches. It was not coloured black, probably because then the number of the Troglodytes was not so great, consequently the remains of their feasts were less, and the products of their decomposition were partly dissolved by the water and washed away underground.

The bed below this red relic-bed consists of a regular bed of *yellow loam*: towards the eastern side it is of considerable thickness; but the depth has not been ascertained, as the water came in when any deep excavations were made. It does not extend over the whole area of the cave, so that in the back portion of it the red relic-bed rests immediately on the original rock. So far as this bed of loam has been examined not the slightest trace of a bone was found in it, though on the surface some few bones and implements, and also some flint-flakes, lay pressed into it, and this fact seems to prove that man was the first occupier of the cave. All the beds were somewhat sloping; the greatest inclination was from the southern entrance to the middle of the cave: the angle of inclination was here about 25 degrees. In front the beds terminate abruptly.

In the process of excavation one bed was carefully taken away

from the others in order to keep the contents perfectly distinct.
(What has been said in several journals, that too little care was
taken in the excavation, is absolutely incorrect.) At first I
considered the two relic-beds as the products of two different
ages; but the determination of the bones found in both beds
(as decided by Professor Rütimeyer when he first examined the
two separate collections of bones), and also the implements,
unquestionably disprove my idea; so that we had here only
one relic-bed, divided simply by colour, which had been
gradually acquired in the course of ages. The whole thickness
of this bed, considered as one, varied from 30 inches to 6½
inches—a fact of very considerable importance. The cubical
contents amounted to nearly 3,530 cubical feet, so that the
average thickness throughout the cave may be taken as 1 foot
9 inches.

In the excavation the greatest possible care was taken, and
the work was narrowly watched. It required every exertion of
mind and body; but this was amply repaid by a most valuable
and interesting collection of objects. It was most fortunate
that the weather was very fine during the excavation. The
greatest difficulty we had to contend with was the quantity of
water which came in—in fact when excavating just by the
eastern entrance we were at least 35 inches under the level of
the water at that season. Much time was required to clear out
the water; pumping had to be carried on for several days with
more or less intermission. The specimens found were carefully
cleaned in an adjacent brook immediately after they were dug
up. The bones were placed in a wicker-work basket, and this,
with its contents, was put into the running water, which partially cleared them from the rich earth adhering to them.
After some time all the specimens were submitted to a regular
cleansing process, and water was poured upon them from a
bucket, till all the mud was entirely cleared off. Neither
brushes nor napkins were used, in order to avoid any chance of
damage to the specimens. After being completely cleaned, the
bones and implements were dried and dipped in a thin solution
of isinglass, which gave an extremely thin coating, but one
which preserved the specimens from further decay. Isinglass
is very much better than common glue, as it is less liable to
grow mouldy.

The *remains of bones* were very numerous. I do not exaggerate
in estimating their weight as 30 cwt. Without any exception,
every bone which contained marrow was broken in piece;

there was not a single bone of the extremities left entire in the whole relic-bed, neither was there a single complete skeleton, nor in fact any portion of one. Though the bone-fragments were so very numerous. I could not find a single instance where the blow appeared to have been given by a pointed or chisel-shaped implement ; but, on the other hand, all the well-preserved specimens, showing where the blow had been given, indicate the effect of a blunt implement. It is very evident that the bones were broken for the especial purpose of obtaining the marrow. The fact that the Esquimaux of the present day make use of marrow as an article of food leads us to believe that the ancient cave-dwellers had the same custom. All the bones adhere more or less to the tongue, a characteristic mark, according to Lyell, of the fossil bones of the quaternary formation. Not a single bone showed any trace of having been gnawed,[1] so that we may conclude that man was not only, as before mentioned, the first, but also the only occupier of the Kesslerloch.

Professor Rütimeyer has submitted all the bones to a careful examination, and has determined them, for which my best thanks are due to him. If, therefore, in the following pages I make mention of the species, or their number, I must be understood to use the information derived from this well-known man of science.

The only representatives of the family *Solidungula (Solipedia)* found in the cave are those of the horse. But of this animal we meet with a number of teeth, vertebrae, scapulae, bones of the feet, and hoofs. From the number of front teeth of the under-jaw 17 + 4 milk-teeth, and of the upper-jaw 15 + 3 milk-teeth, we may calculate that the number of horses eaten in the cave may probably have been about twenty, of which one-fifth had been young animals. From the size of the teeth, the hoofs, and the other bones, we may conclude that the horse of that period was about as large as that of the present day, which only differs from the cave-horse by having a wider foot. Some very few teeth, however, probably show a slight indication of another race, which has much resemblance to the fossil form of the horse found in the quaternary formations.[2] Whether these

---

[1] See the note as to gnawed bones under the description of the Carnivora of the cave.

[2] If it be lawful for an unpractised observer to express any opinion at all on a matter like this, it may be mentioned that, on comparing the few horse's teeth I was able to procure from the rubbish-heap with those drawn in Professor Owen's beautiful plate in the *Philosophical Transactions* for 1869, Plate LVII., most of them appear to

horses lived wild or as domestic animals naturally cannot be made out from the bones, for though by taming or domestication the race may be improved, it is uncommonly difficult, if not impossible to recognise in the bones such an amelioration. The fact that amongst the number eaten one-fifth were foals would lead us to infer that these animals were wild; for if they had been kept as domestic animals for any particular service, they would certainly not have been killed and eaten when quite young. We need not think that horses were domesticated for the sake of their flesh, for it is evident that at that period nature had given to man abundance of flesh to eat. In fact, on the whole, we have not one single reason for believing that horses were domesticated for food; and at the present day we find the horse running wild, though not in any great numbers, as, for instance, on the high steppes and plateaux of Asia and Africa, where small herds of wild horses are known to roam. There can be no doubt whatever that the number of those horses which have become wild is much greater—that is, those which at one time were domesticated, and afterwards escaped from the dominion of man. Our cave cannot show a single trace of these Asiatic or African wild horses of the present day. What is found here is the genuine *Equus caballus*.

The bones of the reindeer (*Cervus tarandus*) were those found in the greatest abundance. At least ninety per cent. of the whole belonged to this animal,[1] as besides a great number of worked and unworked antlers, very many vertebræ, broken bones of the extremities and those of the foot and the toe were met with, and more particularly, a great many broken pieces of the under-jaw, together with the teeth. In order to ascertain the number of reindeer which had been killed here, the lower undermost molars from the right side, and also those from the left side, were selected and counted, as by their form they are easily distinguished from the other molars; they numbered 200 lower hindermost molars from the right side, together with 48 milk-teeth of the same kind, and 180 lower hindermost molars from the left side, together with 48 milk-teeth, so that at least 200

correspond almost exactly with those drawn in figs. 1 and 2, which are the teeth of *Equus caballus*. At the same time there is, as mentioned in the text, a very slight approximation in one or two to those drawn, figs. 6, 7, and 8, which are those of *Equus spelæus*. It is only, however, for regular comparative anatomists to judge in matters of such nicety as this.

[1] Professor Rütimeyer in his late work estimates the quantity of reindeer bones and horns as ninety per cent. of the whole mass; but in the number of individuals the Alpine hare preponderated over the reindeer. *Die Veränderungen, &c.,* p. 16.

grown animals and about 50 young ones, or in all 250 reindeer had been slaughtered and become food for the inhabitants of the Kesslerloch. It may easily be imagined that this was not the whole number of the animals destroyed, for it is hardly to be supposed that every meal was taken inside of the cave. Very probably these primitive hunters consumed a great many in the open air on their hunting expeditions; it was therefore more or less a chance that so many reindeer bones were found in the cave.

At the present time the reindeer does not live in our latitudes, but has withdrawn to the extreme north of the Old World, and if the American reindeer may be considered the same species as ours, to the corresponding parts of the New World also. It lives in all countries north of 60 degrees, but in many places it is found occasionally up to 52 degrees of north latitude. Thus it is restricted by the climate to a band of about thirteen degrees in the north. From this fact it may be safely concluded that the climate of our district at that time must have been much colder than it is at present. The bones of the reindeer of that age do not differ in the smallest degree from those of the animals living wild at the present day, so that we may conclude with tolerable certainty that the reindeer lived wild here as well as in Veyrier, France, and Belgium. Cæsar gives the following account of this animal :—'In the Hercynian forest there is an ox in the form of a stag, and from the middle of the forehead between the ears there arises a single horn, higher and more straight than those horns which are known to us, the top of which spreads out widely, like a hand, or like branches of trees. The female is just the same, both in appearance and the size of her horns.'[1] According to the view of Mr. C. Vogt, the absence of the domestic dog indicates the wild condition of the reindeer. In fact it is difficult to conceive how herds of these animals could have been preserved by man without having careful and watchful assistants well accustomed to running and jumping.

Another animal with divided hoofs is the stag (*Cervus elaphus*),

---

[1] *Lib.* vi. 26. The German version seems to have been a free translation; the quotation as given above is as nearly as possible in the very words of Cæsar. The author seems to have no doubt whatever that the reindeer was intended, but his conviction does not appear to be shared by *all* the commentators. Probably it will be better to give the quotation in the original : 'Est bos cervi figura, cujus a media fronte inter aures unum cornu exsistit, excelsius magisque directum his quæ nobis nota sunt cornibus. Ab ejus summo sicut palmæ ramique late diffunduntur. Eadem est fœminæ marisque natura, eadem forma magnitudoque cornuum.'

the remains of which are met with but sparingly in the cave of Thayngen; the few teeth and bones which have been found there indicate only about half a dozen individuals. The size of the bones is striking, for they approach in dimensions nearer to those of the wapiti or *Cervus canadensis*, than those of the red stag of the present day. Consequently we meet here with exactly the same phenomenon as at Veyrier, at the Robbers' cave at Schelmengraben, and in most of the lake-dwellings. As the investigations of the caves of Belgium by Dupont have proved the existence at that period of this greatest of living stags, it may not, as it appears, be very improbable that some of these larger bones may have belonged to the *Cervus canadensis*, which now wanders about in North America. It would certainly be of great interest if this phenomenon were confirmed by other discoveries. In early ages the red stag abounded in our country and Switzerland, as may be proved by the remains of bones in the lake-dwellings, where they are found in great quantities. But at the present time this noble beast is not to be found in our country, and it is of very rare occurrence in a great part of Germany. It is still abundant in Poland, Austria, and in the Caucasus, so that it seems limited to a breadth of land between the 45th and 65th degrees of north latitude. Consequently it overlaps (so to speak) the limits of the reindeer, and therefore exists at the present day together with it, and yet it prefers a temperate rather than a cold climate. From the fact that the red stag's remains are met with very sparingly in the Kesslerloch, while on the contrary they are abundant in the lake-dwellings, we may conclude that our climate had become milder during that interval of time. The only reason why the animal is not living amongst us at the present day is that it has been very severely hunted, so that it has become peculiarly shy and fearful.

The remains of the urus, ur, or *Bos primigenius*, were still more rare than those of the stag; only some small bones were found, chiefly those of the feet, and fragments of the skull and the medullary bones. The animal is extinct, but the remains of it found in England, Scotland, in the Danish kitchen-middens, in Sweden, France, Belgium, Germany, Italy, and Austria, prove to a certainty that it must have spread over the whole of Europe. It is not so very long since it disappeared from the face of the earth. Even in Cæsar's time the urus was to be found in the Hercynian forests, and consequently it lived there at the same time as the reindeer. He describes these

animals in the following words. 'In size they are little less
than elephants; they are like bulls in appearance, colour, and
shape. Their strength and speed are great, and when they see
either man or beast they rush to the attack. They (the natives)
catch them in pit-falls made with great care, and then kill them.
The young men are hardened by this laborious work, and are
accustomed to this kind of hunting, and those of them who
have killed the greatest number are considered worthy of the
highest honour; the horns being exhibited in public as a tes-
timony of their prowess. The uri, even when taken very
young, cannot be tamed or domesticated. In size, form, and
appearance the horns differ much from those of our cattle.
They are carefully edged with silver, and used as drinking-cups
in their greatest feasts.'[1] The 'Niebelungenlied,' which dates
from the 12th century, mentions this animal, for it says of
Siegfrid:

> Dar nâch sluoc er schiere
> Einen wisen und ein elch
> Starker ûre viere
> Und einen grimmen schelch.—880. Lachmann's edition.

Which may be thus translated, 'On which he quickly slew a
wisent and an elk, four strong uri, and a furious schelch.'[2]

It has moreover been proved that this animal continued in
a wild condition in Europe till the middle of the 16th century,
nay, in England even a century later. It has often been affirmed
that the *Bos primigenius* is the ancestor of our race of domestic
cattle; there are, however, many opponents to this view.
Professor Rütimeyer, who is acknowledged as one of the first
authorities in this department, in his treatise on the Animal
Remains of the Swiss Lake-dwellings, gives it as his opinion that

[1] *Lib.* vi. 28. The quotation as given by the author being only a portion of this interesting passage, a translation of the whole has been substituted in the text for the few words he gives, and it may be as well also to give the original. 'Tertium est genus eorum qui uri appellantur. Hi sunt magnitudine paullo infra elephantos; specie et colore et figura tauri. Magna vis eorum est et magna velocitas, neque homini neque ferae quam conspexerunt parcunt. Hos studiose foveis captos interficiunt. Hoc se labore durant adolescentes atque hoc genere venationis exercent; et qui plurimos ex his interfecerunt, relatis in publicum cornibus quae sunt testimonio magnam ferunt laudem. Sed adsuescere ad homines et mansuefieri ne parvuli quidem excepti possunt. Amplitudo cornuum et figura et species multum a nostrorum boum cornibus differt. Haec studiose conquisita ab labris argento circumcludunt atque in amplissimis epulis pro poculis utuntur.'

[2] The word 'Schelch' has given occasion to many discussions and opinions, which it will be needless here to recapitulate, as it appears to me that no argument has been brought forward to prove positively what the animal was. Giant stag, buck, megaceros, and reindeer are some of the versions given by different authors.

there is much probability in the supposition, that our Swiss domestic cattle were derived immediately from the tame cattle of the lake-dwellings, with which the modern race of great cattle in the Swiss valleys have far more resemblance than they have with the uri. The urus was at least about a quarter larger than a great cow. The horns sprang with a broad base in front of the forehead, then bent very decidedly backwards and outwards, and then sharply forwards and upwards.

Nearly allied to the *Bos primigenius* is the Wisent, Aurochs, (*Bison priscus*, or *Bos bison*). Of this animal we have found a considerable number of bones of the feet, some teeth both of old and young animals, a hip-bone and humerus, a large number of pieces of broken bone, a few fragments of the skull, and two horn-cores, which measure a foot both in length and in their greatest circumference. The remains of this animal were consequently much more numerous than those of the allied species last mentioned, as we may calculate the number of individuals to have amounted to six. At the present day the *Bison priscus* is the largest mammalian in Europe, for it is 7 feet high, 13 feet long, and attains a weight of from 18 to 20 cwt. The district which it inhabits in Europe is very limited : for it occurs in Russia, but in fact only in the forest of Bialowies, where it is under the special protection of the Russian Emperor, to whom we are indebted for having preserved this animal from extinction. In the earlier ages it was much more widely distributed, and was found not only over a great part of Europe and Asia, but also over North America. In the eighteenth century it was to be found in Transylvania. For some ages it was living even in Switzerland, as may be proved by the remains found in the deposits of gravel and in the lake-dwellings. Eckehard mentions its occurrence as a wild animal near St. Gall in the year 1000. In England and Scandinavia, and generally in the far north, it has been long extinct, so that there it belongs to prehistoric ages. It is distinguished from the *Bos primigenius* by the hump on the fore part of its back, and by its horns, which are somewhat small, and at first bent outwards and downwards, and then upwards and forwards.

A third kind of ox has also been found, namely, the tame ox, or *Bos taurus*, of which, however, only two phalanges have occurred, and these seem to be identical with those of the so-called marsh-cow of the lake-dwellings. Though these bones show no difference either in colour or state of preservation from the others, yet I cannot but think that they do not belong to

the same period. If it be true that these Troglodytes kept
domestic cattle, how is it that we have but two phalanges? In
all probability these bones, when the upper covering was removed,
had inadvertently got mixed with those from the relic-bed, and
if so, are of the age of the lake-dwellings, for that the human
representatives of this latter period knew of this cave and
occupied it, at any rate temporarily, is shown by the fragment of
pottery found in the uppermost covering (Plate XIII. fig. 80),
which is made of the same material as the pottery found in the
lake-dwellings. Consequently it is not improbable that these
bones may be comparatively modern, and may date from the
age of the lake-dwellings. Or is it possible, in case these two
phalanges really came from the relic-bed, that they may have
belonged to the musk-ox or sheep, which is well known to be
one of the smallest of the tribe, and which lives in the dreary
waste of Tundra, and consequently is the genuine representa-
tive of a cold climate? A piece of sculpture, which will shortly
be described, gives some grounds for this idea.

The two stately and beautiful inhabitants of the Alps, the
chamois (*Capella rupicapra*), and the ibex (*Capra ibex*), have left
in the cave some very few remains as proofs of their having
formerly inhabited the neighbourhood of Thayngen. Of one of
these animals there was found the horn-core, and of the other
probably a dozen teeth and a humerus. Both are well-known
genuine types of the dwellers in our highest mountains; they
seldom descend to the lower hills or the plains, probably from
fear of the hunter, and perhaps from the nature of the climate.
From the very few remains of these two animals in the cave, I
can hardly, for my part, conclude that they were rarities to
these Troglodytes. It is rather to be supposed that at that time
they were found in tolerable plenty on the Jura Mountains of
the borders and their spurs, but, on account of their rapid
flight to the very highest points, they were difficult to secure;
it is probably on this account that we find these animals so rare
even at the foot of the Alps, in the heights at Salève, and Villeneuve. It will not be long before the ibex will be an extinct
animal; in fact it is no longer found in Switzerland. It is
probably not the climate which has driven it away from our
borders, but the fact of its being incessantly hunted by man.
At all events, the presence of both these animals remains in the
cave of Thayngen indicates an Alpine climate in the district at
that period.

The order of Pachydermata is represented by the mammoth

(*Elephas primigenius*) and the *Rhinoceros tichorinus*. A considerable number of bones of the mammoth have been found, amongst which are several portions of the skulls and various bones of young animals of different ages, phalanges of grown elephants, and a large number of broken fragments of great bones. Most of these elephant's bones come from the lower relic-bed; some of them lay immediately upon the clay; most of them, like many remains of the horse, the glutton, and the wolf, have a coating of tufa. But that the bones of this pachydermatous animal do not belong exclusively to the lower beds, but also to the higher or black relic-bed, is proved by the discovery in the latter bed of two molars, a tusk, and a great many fragments probably derived from this very tusk. One of these molars is entirely broken to pieces, and we only secured two plates, while the other is perfect down to the roots. It shows plainly nine plates, and therefore belonged to a young animal. The tusk was found near the pillar hardly 2½ inches below the top of the uppermost relic-bed. It was 53 inches long, and weighed 42 pounds; but it was so much decayed that in spite of every care, and though we bound it round with narrow strips of cloth, it fell to pieces. At first I had the idea that the remains of this prehistoric animal had been found in the alluvial deposits, as they are occasionally found in the gravel of the Rhine, and that they had been brought into the cave by the inhabitants as curiosities. But further investigation has brought to light a number of other elephant's bones, molar teeth, and tusks both of old and young animals, so that it is certain that these animals have lived here, been hunted and taken into the cave. For I see no reason why the inhabitants should have brought in and broken to pieces the useless bones of this animal, which they had found by chance; nor why so many young animals should here have met with their death, unless it had been caused intentionally by the hand of man. The Kesslerloch thus proves afresh the contemporaneity of man and the mammoth. When it became extinct we know not, for its disappearance from the face of the earth must in any case be thrown back to prehistoric times. Thousands of years ago this giant animal was very widely distributed. Its remains are found not only in the whole of North America, from Behring's Straits down to Texas, but also in the Old World, from the extreme north of Siberia to the furthest parts of Western Europe. In fact it crossed the mountains, and ventured even up to 37° north latitude. Fortunately we have not merely the bones of this colossus, but we have its whole body preserved under the frozen soil of Siberia, where it is well known to occur in such abundance that a regular trade is carried on with the ivory from this fossil animal. It differs from its descendant, the modern elephant, by being provided with hair: the dark grey skin was covered with reddish wool, mixed with long black bristles somewhat thicker than horsehair. A thick mane hung from the neck, as is shown by the drawing or etching found in the cave of La Madeleine.[1] Nature had therefore made every provision to protect this animal from cold. Its presence in the Kesslerloch is a fresh confirmation of the view we have already stated, that at that period we must have had a regular Arctic climate. The food of this giant animal consisted of the leaves of the fir, as is shown by the remains found in Siberia, both in the intestines and between the teeth. As mammoth bones have been found together with those of the horse and the reindeer, as well as both below and above them, the excavation at the Thayngen cave proves that these two last-mentioned animals lived together with the mammoth.

This animal was constantly accompanied by the *Rhinoceros tichorinus*, which has already been mentioned, the remains, of which, however, in the cave were very scanty: they consist only of three teeth, probably of an old animal, and of some few pieces of the skull. It is called *tichorinus*, as it had a bony plate dividing the nose and supporting the horn, which was three feet long; this peculiarity distinguishes it from all the seven species living at the present day. The extent of the distribution of this animal was about the same as that of the mammoth, and, like it also, it was protected from the cold climate by thick wool. About a hundred years ago, in the 64th degree of north latitude, on the banks of the Wilni, a specimen of this beast was found preserved in ice, and almost entirely perfect. It is a striking fact that the species most nearly allied to these two furred pachydermata occur in warm districts, and have nearly a naked skin.

The order of Carnivora is very well represented. It may easily be imagined that the wolf (*Canis lupus*) is not absent. We have found a large number of different parts of the skeleton, eight right and seventeen left under-jaws, and five right and three left upper-jaws, so that the whole number of wolves brought

---

[1] See *Reliquiæ Aquitanicæ*, part xiv., B: Plate XXVIII.

here may amount probably to seventeen. Neither their teeth
nor their bones are at all different from those of the present
wolf, so that we may conclude that this veteran, of which so
much has been said in the oldest histories and legends, has
remained unchanged for thousands of years. It is one of the
few animals which are able to adapt themselves to all varieties
of climate. It has now withdrawn from this peaceful valley,
never more to return to it. And yet it is widely distributed.
It has become rare in Switzerland. Evidently its flesh was
eaten by the cave-dwellers of the Kesslerloch, as well as that
of all the animals found there, for all its bones were broken
in pieces, like those of other animals, which certainly would
not have been the case if the flesh of the wolf had not been
used for food. Without doubt the wolf was not a welcome
visitor to our Troglodytes : not only their safety, but even their
very existence was endangered by its presence ; for the wolf and
the glutton were the most deadly enemies of the reindeer, to
which the cave-dwellers owed so much, for, as with the modern
Esquimaux, it furnished them with both food and clothing.

The fox tribe was not wanting, for the remains of *Canis
vulpes, Canis fulvus,* and *Canis lagopus* were met with. Of the
first only a very few teeth were found—at the most two or three ;
of the second ninety pieces of the under-jaw, representing forty-
five to fifty individuals ; and of the last forty-six under-jaws
from at least twenty to thirty individuals, so that altogether
there were about eighty animals. The upper-jaws and other
parts of the skeleton of all three species were found in the
cave, but singularly enough, very rarely. So that the common
fox at that time was not very numerous in our district, but
the remains of the Arctic fox and the North American red
fox are found in greater abundance. The first at the present
day is known only in the most northern parts of the Old and
the New World, where the ice spreads out in enormous plains.
It has consequently entirely withdrawn from the modern fauna
of Thayngen, and this probably is the most striking argument
in favour of the former rigour of the climate. *Canis fulvus* is
no longer found in Europe, but only in the northern parts
of North America. There is considerable doubt whether the
domestic dog (*Canis familiaris*) was an occupier of the Kesslerloch or not. Its presence here would, as before remarked,
lead us with great probability to conclude that both the
reindeer and the horse were also domesticated. Professor
Rütimeyer himself says that it cannot be positively decided

whether a questionable upper-jaw and tibia belong to the
dog or not. Both these bones indicate an animal of the size
of the Esquimaux dog, or of that of the American prairie-wolf
(*Canis latrans*), which is well known to be smaller than our wolf;
and as we have no more certain evidence to go upon, we must
come to the conclusion that the dog at that time was not the
faithful companion of our Troglodytes, for had it been so,
and had he as a domestic animal been used for the chase,
most certainly his remains would have been found in greater
abundance. And, again, the fact of no bones having been
found which had been gnawed argues against the existence of
the dog at that period.[1]

The family of cats is also represented amongst the remains.
We find the wild cat (*Felis catus*), of which however only one
under-jaw of a very large animal has come down to us, and
yet at the present time it is nearly spread over the whole of
Europe, but is not found beyond this continent. Mountain
forests are its favourite haunts. Most singularly, it is not
found in Norway, Sweden, or Russia.

In the countries just mentioned the wild cat is replaced by
the lynx (*Felis lynx*), and the remains of this animal are also
met with in the Thayngen cave; three right and one left
under-jaw and one upper-jaw have been found. At the
present day it is widely spread over the above-named countries,
and generally over the whole of Northern Europe. Even in
Switzerland two or three individuals are killed every year.

Amongst this varied series of animals the appearance of
the cave-lion (*Felis spelæa*) is of no little interest. Its existence
is shown by two canines of a full-grown animal—a right
upper-jaw, a right upper tusk, a tooth of the left under-jaw and
one of the right under-jaw, and an upper canine tooth of young
animals. The cave-lion is now extinct. On the average it was
larger than the lion of the present day. As yet it has been found
only in French, Belgian (English,[2]) and Swabian caves; also in

---

[1] Amongst the specimens which came from the rubbish-heap of the cave when it
was resifted were several pieces of bone, some of which I secured and brought to
England. On showing them to my friend Mr. Pengelly, F.R.S., the well-known
explorer of Kent's Cavern, he at once pronounced that one of them *has* been gnawed;
and, since then to prevent any mistake, his attention has been again drawn to this
specimen, and he quite adheres to his previous opinion. No one is better able to decide
this question than Mr. Pengelly.

[2] It hardly needs an apology for inserting here the word 'English,' which is *not* in
the original. It is singular that such a slip could be made by the author, as a full
notice of this animal's remains in England was published by Messrs. W. Boyd-Dawkins
and Ayshford Sanford in the Palæontographical Society's publications in 1868.

Italy and Sicily, and perhaps also at Natchez, in Mississippi.
It is well known that the modern lion now inhabits only warm
countries; it is therefore a matter of surprise that its pre-
decessor should have been found associated with animals which
are never met with in warm countries. It must be remembered,
however, that a species nearly allied to the lion, the royal
tiger, in the course of its pursuit after prey presses forward
even to the 52nd degree of north latitude, and that the larger
species of cats are generally very easily acclimatised, and thus
what was at the first moment surprising falls to the ground.
The lion inhabited Europe in historic times, for Herodotus
tells us that the camels belonging to the army of Xerxes
were attacked by lions in Thessaly.

Four right and two left under-jaws and two fragments of the
upper-jaw of the *Gulo luscus*, or glutton, were found in the cave,
one of which came from the lowest bed of the red relic-bed; and
this may be looked upon as a proof of the great antiquity of this
animal. At the present time it lives only in the northern
parts of the Old and New World, while in earlier ages it was
found much further south. It was the deadliest enemy of the
reindeer, so that it may be taken for granted that in the pursuit of its prey it wandered together with it towards the
northern regions. Its appearance in the neighbourhood of the
cave at Thayngen would not create a pleasant sensation in
the minds of its inhabitants.

Even the bear (*Ursus arctos*) was not wanting amongst this
mixed company. We have however found only a very few
teeth, and parts of the skeleton, and one almost perfect upper-
jaw. We can only trace the parts belonging to two or three
individuals. The rapid advance of civilisation has very much
limited the spread of this animal, so that it will soon have to be
considered as an exceptional straggler in Europe. It is known
that the bear is occasionally met with in the Alps of Wallis,
and the Grisons. In early times it was widely distributed in
Europe. Its remains are found in the Swabian caves together
with those of the cave-bear, of which there is not the smallest
trace in the Kesslerloch. Yet the common bear at that time
was of rare occurrence, as most of the cave 'finds' show; it
appears not to have been found in any great abundance till the
cave-bear, which was much greater and stronger, had given
place to it.

The Rodentia were another order of animals found in the cave.
A radius of the marmot (*Arctomys marmota*) and innumer-

able bones of the Alpine hare (*Lepus variabilis*) were discovered here. The remains of the hare were by far more numerous than those of any other animal. Not less than 124 right, and 502 left under-jaws were found, so that more than 500 of these animals have been brought into the cave. At the present day the Alpine hare no longer occurs in this district, but has retreated to the higher regions of the Alps, where it is still found in as great abundance as the field-hare in the plains. Of this last-named animal some few remains of bones have been found in the cave; but they are doubtless of later ages.

The remains of several birds were found in the cave of Thayngen; we may mention in the first place no inconsiderable number of bones of the ptarmigan (*Tetrao Lagopus*). Not less than eighty right, and the same number of left humeri were found, so that the cave-dwellers had eaten at least eighty individuals of this species. A great number of other bones were found, but not the trace of a skull. This appears the more singular, as the examination of the cave at Veyrier has given precisely the same results. Naturally we cannot come to any certain and definite conclusion respecting this fact. Professor Fraas is disposed to attribute a number of these bird's bones to the *Tetrao albinos*. The ptarmigan is found wherever the mountains are covered with snow and ice, consequently it lives both in our Alps and in the cold north, and it is in fact the only animal of the Thayngen fauna which has withdrawn from this district both to the greatest heights of the south and to the most distant districts of the north.

The goose was also found together with the ptarmigan; but it cannot be decided whether it was the *Anser cinereus*, or *A. legatum*, as only the heads of three humeri were found, two of which had been worked; the humerus had been cut off about the middle, and then a little distance off it had been perforated.

Of the wild swan (*Cygnus musicus*), we found the heads of two humeri and a tibia, which had been cut with a flint knife. It is to be found in every region of the world except the torrid zone; and yet it is more abundant in the temperate and cold districts of the northern hemisphere. Its favourite haunts are fresh-water lakes and watery marshes, and these were doubtless found in those ages in the immediate neighbourhood of the cave. The peat moors lying east and west of the Kesslerloch, and the lake of Egel, which still exists, are perhaps the last traces of these marshes.

Six humeri and one tibia of the raven were found in the cave.

The only remains of a predaceous bird met with were those of the sea-eagle (*Haliaëtus albicilla*). It is a very widely spread bird, and is found in the whole of Europe, in a great part of Asia, and in all the rivers of North and West Africa. One or two bones of the feet are all we have to prove its existence.

We may probably consider as chance introductions into the cave, a number of vertebræ and pieces of the skull of probably the common snake, and bones of the shrew mouse and the frog. [1]

Such is a rapid review of the Thayngen fauna at the time when the Kesslerloch was inhabited by man; for general reference I will add a table of the representatives of the animal world found here, with a note as to their number, and also as to the direction in which they have retreated from the neighbourhood of Thayngen.

Immediately after the foregoing sketch had been printed, two pieces of bone were found amongst the rubbish; one of them was that of a bison; the other probably belonged to a rhinoceros. On the first is scratched or engraved the figure of a fox: and on the second that of a bear. Both figures are drawn sitting, and from being represented so naturally, form a fitting appendage to the reindeer and the horse. The mode, however, in which the work is carried out betrays the hand of an unpractised artist, for these drawings want the nicety and correctness which the others possess in the highest degree. —See Plate XV., figs. 98 and 99.

Even a superficial glance at this list of animals is sufficient for the conviction that the fauna of those ages was totally different from that of the present day. Nowhere else in the whole world do we find the remains of these animals so associated together. Some of them, such as the cave-lion, the mammoth, the rhinoceros, and the *Bos primigenius*, have long since disappeared from the earth's surface: for it is clear that they had on the one hand to struggle for existence with the human race, which was far superior to them in intelligence and cunning, and on the other they had to succumb to the changes of climate. For all these animals, with perhaps a single

---

[1] Professor Rütimeyer seems to consider the bones of the snake and of the shrew-mouse to be late introductions by chance into the cave, and he says that probably the same ought to be said of the European fox, on account of the very great rarity of its remains. *Veröadenougen*, &c. page 44.

			No.	Animal	Count	
EXTINCT ANIMALS			1.	Cave-lion	3	individuals
			2.	Mammoth	4-6	,,
			3.	Rhinoceros	1-2	,,
ANIMALS WITHDRAWN FROM THE DISTRICT	To the North		4.	Urus	1	,,
			5.	Reindeer	250	,,
			6.	Glutton	4	
			7.	Arctic fox	3	,,
	To the Alps		8.	Chamois	1	,,
			9.	Ibex	1	,,
			10.	Alpine hare	500	,,
			11.	Marmot .	1	,,
	To America		12.	Wapiti (?)	1	,,
			13.	Red fox .	40-50	,,
	To the North and the Alps		14.	Ptarmigan	80	,,
	To the adjoining districts and countries		15.	Wiscat .	6	,,
			16.	Red-stag	6	
			17.	Bear	2-3	,,
			18.	Lynx	3	,,
			19.	Wild cat	1	,,
			20.	Wolf	17	,,
			21.	Wild swan	1	,,
			22.	Wild goose	2	,,
			23.	Sea-eagle	1	,,
	Still living in the district		24.	Common fox .	2-3	,,
			25.	Field hare (?)	2	,,
			26.	Raven	3	,,
	domestic animals		27.	Dog (?)	1	
			28.	Horse	20	,,

exception, are genuine types of animals belonging to a very cold
climate. Again, other animals, such as the reindeer, Arctic
fox, red fox, glutton. ptarmigan, and the wapiti, have retreated
to the far north. The chamois, the ibex, the Alpine hare,
and the marmot have wandered off to the Alps, where it
is evident the colder climate agrees better with them than
the temperate climate of Middle Europe. It would probably be correct if we take for granted that these animals, not
only at the present day, but at all periods, were accustomed
to a cold climate, and further, that wherever their existence at
any time can be proved, as for instance is the case in the
Kesslerloch, the argument in favour of a former cold climate is
conclusive. Even the remaining animals, the wolf, the fox,
the wild cat, the lynx, the bear, the wild goose, the wild swan,
and the sea-eagle, all prefer a cold climate, or at any rate are
well able to bear it. An Alpine climate is not injurious even
to the deer and the horse. There can therefore be no doubt
whatever that the neighbourhood of Thayngen had formerly,
in the strictest sense of the word, and for a lengthened period,
an *arctic-alpine* climate, which, however, from various causes
became gradually considerably milder ; and this gave the above-
named animals the opportunity of migrating to the north and
the south, so that of the twenty-five animals which have been
determined with certainty, only two species now live in the
district, namely, the fox and the raven. What a surprising
fact !

It may be interesting to compare the fauna met with in the
Kesslerloch with that of some other abodes of prehistoric man.
For this purpose I have prepared the following table, which
however makes no pretence of being perfectly correct. It
seemed proper here only to insert the characteristic forms of
animals, and not the actual species, which owe their existence
to local conditions. The following is the meaning of the marks
given in the table : A single asterisk indicates a single individual, two asterisks several individuals, three asterisks indicate
a large number, and four asterisks a very large number of the
animals mentioned.

Animals	Kesslerloch	Veyrier	Belgian Caves	Aurignac	Swedish Caves	Danish Kitchen-middens	Lakes Dwellings
Horse . . .	***	**	***	***	***	—	*
Reindeer . .	****	****	****	**	***	—	—
Red deer . .	**	**	**	*	**	***	****
Chamois . .	*	*	*	—	—	—	*
Ibex . . .	*	**	**	—	—	—	*
Bison priscus .	**	—	**	***	*	—	*
Bos primigenius .	*	—	*	—	*	**	—
Bos taurus . .	?	*	?	—	—	—	****
Pig . . .	—	**	**	*	**	****	****
Mammoth . .	**	—	**	*	*	—	—
Rhinoceros .	*	—	***	*	*	—	—
Wolf . . .	***	**	**	**	*	**	*
Arctic fox . .	****	—	**	—	—	—	—
Common fox . .	*	**	—	***	*	**	**
Dog . . .	?	—	—	—	*	**	**
Lion . . .	**	—	*	*	*	—	—
Hyæna . .	—	—	**	**	*	—	—
Lynx . . .	**	*	**	—	*	**	—
Wild cat . .	*	—	**	*	*	**	*
Cave-bear . .	—	—	**	**	****	—	—
Brown bear . .	**	*	**	*	*	**	—
Glutton . .	**	—	**	—	—	—	—
Marmot . .	*	**	**	—	—	—	—
Alpine hare . .	****	**	—	—	—	—	—
Field hare . .	*	**	**	—	*	—	—
Roe . . .	—	—	**	**	—	****	**
Badger . .	—	**	**	*	*	—	**
Beaver . .	—	*	*	—	*	**	**
Elk (or moose deer)	—	—	*	—	—	—	**
Ptarmigan . .	****	****	**	—	**	—	—
Wild swan . .	*	—	—	—	—	**	***
Goose . .	*	—	—	—	—	*	***

If we compare the fauna of Thayngen with that of Veyrier, a glance at the foregoing table will lead us to the conclusion that there is much similarity between them. At both localities the reindeer, the horse, and the ptarmigan are the most prevailing species. The small number of Alpine hares at Veyrier is striking as well as the entire absence of the remains of the Arctic fox, the glutton, and the wild cat, while the badger and beaver are wanting at Thayngen. The essential difference between these two faunas consists in the complete absence at Veyrier of the mammoth, rhinoceros, cave-lion, Bos primigenius, and Bison priscus; and this might almost lead to the conclusion that the remains in the cave at Veyrier are of later date than those of the Kesslerloch. The species however common to both localities are the chamois, the ibex, the marmot, the wolf, the common fox, the lynx, the brown bear, the Alpine hare, the reindeer, the horse, and the ptarmigan; an assemblage of animals which, as before remarked, never afterwards lived together, and the occurrence of which together at the same time evidently indicates the same conditions of life.

The similarity of the Thayngen fauna with that of Belgium is still more striking. All the animals found in the Kesslerloch, with the exception of the birds, occur also in Belgium. On the other hand, the Belgian caves have certain characteristic animals, as, for example, the giant stag, the cave-bear, the cave-hyæna, the *Elephas antiquus* and the *Rhinoceros Merkii*, and probably a few more. The prehistoric fauna is consequently richly represented there. But if we consider that the fifty species (or thereabouts) in the Belgian fauna have been procured from thirty-eight different caves, the similarity between the fauna of Belgium and Thayngen, bearing in mind that the latter is in Northern Switzerland, is very striking. This fact leads to the undoubted conclusion that both countries must evidently at one time have had an Arctic climate, which quite agrees with the views of geologists.

Even the fauna of the Swabian and French caves exhibit a great agreement with that of the Kesslerloch, so that from this fact we are led to the conclusion that not only Switzerland and Belgium, but generally the whole of North and Middle Europe, must have had an Arctic climate. The only difference seems to be that at Thayngen, as already remarked, the cave-bear is wanting, while in the Swabian caves it is very numerous, and in some French and Belgian caves it occurs in considerable numbers. The fact of this animal not being found at Kesslerloch is at the first moment somewhat surprising, and might almost lead to the idea that the discoveries in our cave may be of later date than those of the above-named caves. But if we consider that in certain Belgian caves, which, according to Dupont, were inhabited at the age of the mammoth, the cave-bear is not to be found, and moreover that, like the mammoth, the rhinoceros, and the cave-lion, it is not to be found in *all* the Belgian and French caves of the reindeer age, we may from this conclude that our bone remains are in any case as old as the greater proportion of the oldest bones found broken in caves. But still it is not improbable that those caves where the cave-bear is found most abundantly as the chief beast of prey were the oldest caves inhabited. M. Lartet is therefore correct in dividing the time when man lived in caves and used stone as his only tool into four periods, viz., that of the cave-bear, that of the mammoth, that of the reindeer, and that of polished stone ; and according to this, our cave remains would have to be placed chronologically at the end of the second and the beginning of the third period of the stone age.

A considerable and striking difference, however, is found if we compare the fauna of the lake-dwellings with that of the Kesslerloch. The whole of the animals which belong to a cold climate are wanting in the lake-dwellings; we find neither the reindeer nor the mammoth, the rhinoceros, the lion, nor even the lynx, the cave-bear, the brown bear, the glutton, the Arctic fox, the marmot, the hare, nor the ptarmigan. But on the other hand, we find the red stag, the wild boar, and the elk, and of tame animals, the *Bos taurus*, the dog, the sheep, and the goat have now become regularly naturalised. These differences are evidently not merely to be ascribed to chance, but doubtless arise also from the alteration of climate. When men built their abodes in lakes and marshes, the climate was considerably milder, and the fauna consequently approached much nearer to that of our own time. The age of the Troglodytes consequently lies far beyond that of the lake-dwellers, although the fauna of the Kesslerloch shows some slight connection with that of the lake-dwellings by the presence of *Bos primigenius*, *Bison priscus*, and the red stag.

The difference between the fauna of Thayngen and that of the Danish shell-mounds is still greater; for that of the latter comes much nearer the modern fauna, nay, is almost identical with it. The Danish kitchen-middens are consequently of more recent date than the older lake-dwellings, and very much more recent than our cave remains.

It is most singular that not a single human bone has been found in the Kesslerloch, except one collar bone belonging to a young individual. It was certainly at one time given out to the world that the complete skeleton of a child had been found in the cave. This is decidedly true; but the reporter who says that he was an eye-witness when it was taken out of the ground does not appear to have noticed that this skeleton lay hardly two inches and a half below the surface of the upper covering, and consequently did not belong to the prehistoric age, but to the present. And here I will take the opportunity of mentioning a fact. I was informed by three men who worked for a time as labourers when the Baden Railway was made, that in the immediate neighbourhood of the Kesslerloch, when they were blasting the rock, a moderately large cave was exposed, within which a considerable number of human bones had been laid up. I made them show me this questionable place, but, in fact, I

found not the slightest trace of a cave. Even though implicit
faith may not be given to the communications made to me, inas-
much as uneducated people make but little distinction between
human and animal bones, yet it is by no means impossible that
the Troglodytes may have brought their dead here to dispose
of them. At all events, it is certain that our Troglodytes were
no cannibals ; neither did they burn their dead, for if so, we
should certainly have found more human remains in the cave.
What the cave-dwellers of the Kesslerloch may have been in
appearance, whether of large or small stature, whether they
looked intelligent or stupid, it is impossible to say, for want of
evidence. The main thing once and for all is the proof of the
existence of prehistoric man in the Kesslerloch near Thayngen.

Though no human bones were found in the regular relic-bed,
yet the number of specimens showing the handiwork and civilisation of our Troglodytes was by no means small. In the first
place, we may mention an enormous quantity of *flint-flakes*, at
least 12,000 in number, the weight of which may be estimated
at nearly seven cwt. A superficial glance is sufficient to dis-
cern amongst them *three definitely characteristic forms*, which are
drawn (Plate I. figs. 1 and 4 ; and Plate II. figs. 6 and 8). The
most common forms are those represented in figs. 1 and 8.
The length of flints of the first form varies from four inches to
less than two inches, and the breadth from eight to six tenths of
an inch. At one end the flake terminates in a point more or less
decided, while at the other end it is either flattened off level or
is bevelled off. This level flattening at one end was evidently
intentionally brought about by the tool used for a hammer, so
that the flint might be set in a bone handle. The bevelled end
of many of the flakes may possibly have arisen in consequence
of the fracture caused by the use of this instrument. On the
middle of the back there extend, more or less prominently, one
or two, or, more rarely, three sharp ridges, which originate in
the fact that the flint by repeated skilled blows was sharpened
like a knife on both sides ; hence the name of ' flint knives.' The
number of these ridges depends on the number of blows required
to make the flint sharp, and therefore it is evidently due to
chance. The under side of these flint knives is always smooth,
and a trifle arched or curved ; evidently arising from the conch-
oidal fracture of the flint. Flints of this form have very probably been used as arrow- or lance-heads. The second general form
of flints have doubtless been used as boring tools : they are
distinguished from those last mentioned by having one end

decidedly blunt, instead of being chisel-shaped and sharp (Plate I. fig. 4). At all events, I have been able with the greatest ease to bore nearly circular holes in stag's-horn with similar pointed flints. I then made the attempt to bore a circular hole in the corner tooth of a small carnivorous animal, like the perforations in teeth and needles found in the Kesslerloch, and in this experiment I was also successful. In less than a quarter of an hour I had bored a very neat round hole. There can be no doubt that in this operation some of the boring implements were spoiled. The third form of these flints is that drawn, Plate II. fig. 6.[1] It has been decidedly blunted at both ends, probably with the view of not hurting the finger when used. Nearly the whole length of the flint is of equal breadth. Here, again, the number of edges or angles shown on the back varies from one to three. Both of the longer sides are sharpened like a knife, and these cutting edges, like those of the classes before mentioned, have a number of large or small teeth, which evidently have been caused by use. Doubtless these implements were employed exclusively for manufacturing tools of horn or bone. They served as scrapers, knives, and saws, and were absolutely essential to our Troglodytes. The length varies from $3\frac{1}{2}$ inches to $2\frac{1}{4}$ inches, and the breadth from $1\frac{1}{4}$ inch to rather more than $\frac{3}{4}$ inch. The largest specimen is $5\frac{1}{2}$ inches long and $1\frac{1}{3}$ inch broad. Together with these general forms there are, as may be imagined, a great many others, which may be considered more or less as waste implements.

If we compare these flint tools with those from other countries —Belgium, France, Sweden, Denmark, and Australia, we shall find a striking agreement in their forms, which, according to my idea, arises more from the peculiar qualities of the flint than from any common derivation of race. Amongst the mass of flint-flakes there were found a very small number of unworked lumps of flint, and no inconsiderable number of flint ' cores ' (Plate II., fig. 7), from which the augers, knives, and saws had been struck off; they show very distinctly the parts from which the flakes had been taken. This, however, was effected by a well-directed blow, and not by pressure on the flint-core.[2] In

---

[1] This figure appears to agree very nearly with that given by Mr. Evans in his *Stone Implements of Great Britain*, p. 370, fig. 319. It is singular that Mr. Evans, notwithstanding the very great attention he has paid to the subject, evidently hesitates in determining its use.

[2] It appears to me that this assertion is a little too positive; there can be no doubt that the modern Esquimaux do form flint implements by pressure. At the meeting of the British Association at Nottingham in 1866, Admiral Sir Edward Belcher ex-

order to make the piece struck off available, it was subjected to
repeated blows with the hammer or some similar instrument,
till it had become of the requisite form. Even if the art of
striking off these flakes cannot be considered to take a high rank,
yet great dexterity was required to prepare them always in
sufficient number. Our cave-dwellers found the material for
these implements in the immediate neighbourhood; it is still
found in no small quantity on the fields at Lohn, and also
disseminated in the neighbouring rocks.[1] As the diameter of the
largest flint masses in the Thayngen Jura hardly amounts to $2\frac{1}{2}$
inches, our flint knives are decidedly inferior in size to those of
the French. The colour of the stone implements found in the
cave, as well as that of the flint nodules just mentioned, varies
greatly; sometimes it is red, like jasper, sometimes yellow, black,
green, and sometimes white; more rarely variegated in colour,
so that a collection of these little knives of different colours
makes a pleasant impression on the eye of the observer. We
do not find in the Kesslerloch a trace of the stone celts like
those found in the Swiss lake-dwellings and elsewhere. Professor Fraas, in his essays on the history of civilisation, thinks
it is hardly conceivable that the inhabitants of rock caves could
do without stone hammers and stone celts, and goes so far as to
consider it as an accident that they are entirely wanting; but
as they are not found either in the Kesslerloch or in the
Swabian caves, and are also wanting in other places, I am led
to the conclusion that the inhabitants of the Kesslerloch neither
knew nor used this kind of stone implement. It is indeed
hardly conceivable that if they had generally used these implements some few celts should not have been lost or got broken,

hibited the tools which were used by the natives for this purpose. The first process
was to make the flint roughly into shape by a hammer of greenish jasper, not bored,
but which had a groove running round the middle of it, by which it was *tied* with
sinews very securely to a wooden handle. The other tool used is of a kind of ivory, in
shape somewhat like a pistol stock, with a piece of hard reindeer horn taking the place
usually occupied by the barrel, and this was very securely dove-tailed into the stock,
and also tied with sinews put on wet. With this instrument the natives strike off small
portions at a time from the flint, and bring it to the required form. Sir Edward
Belcher stated that he had frequently seen this done, and that he had himself per-
formed the operation. The Indians of California use the same process in making tools
from obsidian. See the *Ancient Stone Implements of Great Britain*, by John Evans,
1872, pp. 34 and 35; and *Reliq. Aquit.* p. 18.

[1] A large number of 'flakes' were found by Mr. Messikomer in the rubbish-heap
from the cave, which he sifted a second time. Many of these flakes were brought
home by me; and my impression, as well as that of my friend Mr. Pengelly, is that
nine out of ten have been derived, either directly or indirectly, from the chalk formation.

in which case they certainly would have been found by us. On the other hand, we have found in the relic-bed about two hundred large rolled stones, the size of the hand, very hard, more or less pointed, all of which show tolerably decided ‘ hammering surfaces ’ and hollows, evidently arising from repeated blows on the bones which contained marrow, and very probably also on the hard flintstones ; so that here in these pebbles we have the most primitive hammers. The well-known proverb, ‘ If men are silent the stones cry out,’ finds here its most literal application.

Although the cave-dwellers were but badly provided with tools, yet on this very account their dexterity in manufacturing implements and ornaments was most remarkable. These, like the tools, show very little variety, and in fact are limited to what was most necessary. The material from which they manufactured their implements was almost exclusively reindeer-horn. No small number of horns were found in our cave heaped together with a still greater quantity of waste from the worked material. The idea of utilising the horns arose probably from their being the only large solid parts of the animal which could be worked up without very great labour. The first thing to be done was to separate the horns from the reindeer after it had been killed. Probably the skull of the animal was beaten to pieces by repeated blows with the rolled stones above mentioned, and then the horns could be separated from it, either entire or with some little portion of the frontal bone (Plate III. fig. 9). Next came the removal of the brow-antler, ending in a palm ; this shoots off immediately above what is called the ‘ burr,’ or rose-piece ; then the first main offset or ‘ bez-antler ’ was taken off, and all this could be accomplished by merely striking off these branches which would have been in the way of working the horn properly. This manipulation was probably not effected without difficulty, as many fragments found in the cave seem to indicate, which have been again and again cut with flint knives in various directions. The stumps or pieces of the branches remaining on the horn were in general very short—hardly more than one or two inches long. Only a single perfect palm was found, while, on the other hand, the fragments were found by hundreds. As the chief offsets or ‘ tynes ’ of a strong antler could apparently be utilised probably as handles for flint implements, they were in general separated from the main antler somewhat more carefully, by making with a flint saw incisions on the opposite sides of the horn down to the porous part, and then breaking

them off. A large number of specimens of this kind were found.
The cuts are often so well defined and sharp, that one is
almost tempted to think that they had been made with a sharp
metal tool. But having tried to cut off in the same manner
a main 'tyne' just as strong with a flint implement chosen for
the purpose, I feel convinced that it is not so, for these cuts were
just as defined as those above mentioned. After the main and
secondary 'tynes' had been taken away, then the main branch
or 'beam' was cut off (Plate III. fig. 11), immediately above
the first main offshoot. This was the last of the preparatory
labours; the raw material was brought into a state to be used
for whatever was intended. The first thing to be done was to
make several incisions with a flint knife lengthwise in the
'beam,' down to the porous part, with the intention of dividing
it into several parts (Plate IV. fig. 12), and then but little had
to be taken away before these pieces became the required implements; this was perhaps the most difficult part of the work.
Every attempt I made to imitate the different parts of the work
succeeded better than this. Immense patience was required to
make a similar incision lengthwise. Several 'beams' were
found with furrows of this kind. One of them (Plate IV. fig. 12)
is about 1½ inch thick, and 16½ inches long, and it has a furrow
9½ inches long and about one-sixth of an inch deep. If this 'beam'
be turned about 60 degrees, a second furrow is seen more than
4 inches long and hardly more than one-twentieth of an inch
deep. Probably the piece of horn chosen did not please the
workman, but was thrown aside as useless. A second 'beam,'
about 15 inches long and 1½ inch wide, has only a single incision,
nearly 12 inches long and one-third of an inch deep. These
longitudinal incisions were made by a sharp-pointed flint, brought
down at right angles on the antler, and then worked till a
furrow was formed. When the flint was so thick as to prevent
the further progress of the work, the furrow was widened into a
wedge-shaped form. All the furrows in the specimens of this
kind now lying before us are consequently broader above (viz.
from ·41 to ·47 decimals of an inch), while at the bottom they
are much narrower (·35 of an inch). This was the manner in
which the horns were in general cut in pieces lengthwise. No
inconsiderable number of specimens show quite clearly the
traces of this mode of proceeding, for they have on both sides
very evidently cut surfaces which in fact were the sides of one
of the furrows. The pieces thus obtained were then scraped
with flint-flakes till they became of the form required. If the

pieces thus cut in two were too broad for an implement, they were again divided in the same manner, but with this difference, that in this latter cut more or less attention was given to the form of the implement required. Thus if it were intended to form an implement with a sharp point, the furrows were made gradually to unite, so as afterwards to save more or less the tedious process of scraping or rubbing down (Plate I. fig. 2). This detailed mode of working the raw material was probably that which was least often used by our cave-dwellers; at all events that mode of working was the more common in which, after the palm, the first main antler, and the points of the upper main 'tynes' had been broken away, the implements required were regularly cut out of the 'beam': at least the majority of the specimens found of this description seem to bear out this view (Plate III. fig. 10). The greater number of the specimens of horn from the rose-piece, the stumps, the palms, and the first brow-antlers show the clearest indications that the main 'beam' was not taken off behind the first main 'tyne,' which, it stands to reason, would save much time, as when the implements were cut off, the main 'beam' was of itself separated from the lower parts of the horn, which would be thrown aside as useless. But even these were occasionally used as implements, for the remaining part which was cut off immediately under the first brow-antler has been found sharpened into a boring instrument. Again, a considerable number of main branches or 'beams' of reindeer horns cut open lengthwise, and with the usual portion of the inside taken away, show that they were cut out to form tools (Plate II. fig. 5). There can be no doubt about the specimen drawn (Plate I. fig. 3) having been thus intended. It is the broken 'beam' of a horn about 1¼ inch thick and 13 inches long, and on the inner or concave side it looks almost like an implement ready made, which may possibly have come into use as an arrow- or lance-head. Doubtless the outline was first scratched in upon the horn, and afterwards it was cut and worked till the porous part was reached, when the piece entirely broke away, and then the part which was hollowed out as above mentioned was thrown aside as useless. It is a striking fact that the inner concave side of the horn was always chosen to be cut out, possibly because this piece of the horn could be better held and more easily worked during the process of manufacture. If the implements made as above mentioned were curved, they were very probably made straight by means of fire, as is done by the Esquimaux of the present day.

D

A glance at the bone implements enables us in a moment to notice certain essential peculiarities, so that we are able to class them into different groups.

In the first group are found all those implements which terminate at one end in an arrow-formed point, and towards the opposite end become regularly broader, and are made sharp on one side, like a chisel (Plate IV. fig. 14). We call these bone implements 'arrow-heads.' They vary much in size, being from $5\frac{1}{2}$ to $2\frac{1}{2}$ inches long, and from $\frac{4}{10}$ of an inch to about $\frac{1}{4}$ of an inch wide. Those found in the greatest abundance were about $2\frac{1}{2}$ inches long and $\frac{1}{4}$ of an inch thick (Plate XIII. fig. 74). It is very singular that the length of the flat cut at the bottom is almost always the same, viz., 1·181 inch dec. About fifty-five of these implements were found, some of which are as perfect as if they had been made but yesterday; the majority, however, were broken. Sometimes the point is wanting, sometimes the side cut, evidently caused by the general use of these implements. In some cases the parallel lines scratched in with flint diagonally across the one-sided cut are still visible. Only one single specimen has similar scratches or furrows on the upper side of the sharpened part. The number of these lines depends entirely on chance; commonly there are five or six. This chisel-shaped end was evidently spliced with the cut surface of a wooden shaft, and the two were bound together by means of a thong of reindeer skin, or platted horsehair, or twisted intestines. The lines on the cut surface were probably only intended to give a firmer hold between the arrow-head and the shaft, so that the work might be stronger. The lines on the upper side gave more hold to the binding material. The bevelled or sloped part is always carried from the under-porous side to the upper and more solid part, and with good reason, for if the contrary had been done the end of the bevelled part would have been formed of the porous and fragile portion, and thus the implement could never have come into use. One of the specimens belonging to this group (Plate V. fig. 19) exhibits on the upper side an ornamentation made of triple parallel lines, and another specimen is ornamented with double intersecting lines lying close behind one another.

The second group of bone implements is composed of those, one end of which gradually tapers to a point, while the other end grows wider, and is sharpened like a chisel on both sides. These we call lance-heads (Plate IV. fig. 15, and Plate V. fig. 18). The bevelled slope is on the average $1\frac{1}{2}$ inch long, and is so arranged that the porous part of the under-side is taken away, so as to guard against the implement being broken when used. It is singular that in this group of implements the sloping cut surface has not a single line or scratch upon it, but, on the contrary, it is in some cases slightly hollowed. The cross cut here is more of an ellipse. They vary greatly in size and thickness, like those of the first group. The largest implement of this kind is broken off at the point, but it is more than $7\frac{1}{2}$ inches long, and very nearly an inch wide; very probably its length originally was twice as much. The smallest lance-heads are about $3\frac{1}{4}$ inches long and $\frac{1}{4}$ of an inch thick. There are all sorts of varieties between the larger and the smaller kinds. One singular specimen is $8\frac{1}{4}$ inches long and about $\frac{9}{10}$ of an inch wide, and on the surface it is regularly burnt, so that one might almost venture to suppose that it had been used as a spit. An exceptional specimen of this group is the lance-head, drawn Plate VI. fig. 26, which has a narrow furrow running lengthwise ·0586 inch dee. wide, and ·0858 inch dee. deep, and on the usual bevelled part there is a round perforation rather more than $\frac{1}{3}$ of an inch in diameter, apparently as a means of hanging up the implement; but whether as a trophy for the owner, who possibly may have made some splendid shot with the weapon, or for some other purpose, it is of course difficult to decide. At all events it is a curious and unique specimen; the length of it is about $4\frac{1}{2}$ inches. The chisel-shaped bevelled part of all these lance-heads must have been intended to be stuck into a split wooden shaft, which was then bound round by some material, and thus the lance-head was made fast. The whole number of lance-heads found in the excavation was about ninety-three, but of these only about a fifth are in good preservation. The larger proportion are only in fragments, but still they all show quite clearly the chisel-shaped bevelling. It may easily be imagined that the points have suffered most; in fact they are wanting in all the broken specimens. Very singularly two fragments of implements of this group have been found made out of a mammoth's tusk. On some specimens of this class, which unfortunately are not quite perfect, the workman has spent more time than usual, and has tried his skill in drawing. Parallel lines grouped in various modes, sometimes on the sides and sometimes on the top, and also zig-zag lines

may be seen as ornamentations on several of these implements (Plate V. figs. 19 and 22).

Nearly allied to these bone implements is another kind of
lance-head (Plate IV. fig. 13), somewhat different from the foregoing. In every case both planes of the chisel-shaped bevel,
and also more especially the sides, have a number of incisions
varying from seventeen to eight, and all drawn parallel
from left to right. In some cases the lines incised cross each
other. There can be no doubt that the intention was simply
first to give the implement more hold in the split portion of
the javelin, and also to secure it firmly in the shaft when
it was drawn out from the body of the animal struck. Another
characteristic distinction consists in the furrows found both
on the upper and the under side. In some specimens they
are only slightly indicated, but in the majority they are very
well marked. Their length naturally depends on the length
of the implement. These furrows commence from $\frac{6}{10}$ of an
inch to $1\frac{3}{4}$ of an inch above the chisel-shaped bevel, and run
on to the point, gradually decreasing in size. They vary from
·157 inch dec. to ·236 inch dec. in width, and from ·236 inch
dec. to ·294 inch dec. in depth. As to the use of these
furrows, there are different opinions; many people consider
them to be channels for poison, while others think them merely
passages for the blood to flow through. There are indeed
savages at the present day who poison their arrows, so as to
make them more deadly in their effect on the human or animal
frame. But I do not believe that they eat the flesh of the
animals killed in this manner; and as the cave-dwellers
of the Kesslerloch, as already mentioned, seem to have fed
upon every animal they killed, it seems to me better to consider these furrows as channels for the flow of blood from
the animals struck, so that they might die more quickly.
It is singular that these furrows and incisions are not to be
seen on all the implements which we have called lance-heads.
But for my part, it seems to me that we must consider these
last-named bone implements as an improvement upon the
earlier ones; and I believe that this view may be proved not
merely by their better workmanship, but also from the fact
that these implements, although certainly occurring together
with those without furrows and incisions, were always found
in the upper relic-bed, and never in the lower. We may
from this conclude that in the later period of the occupation
of the cave by man both kinds of lance-heads were used, while

in the earlier ages only those without furrows and notches
were employed. It is in general very difficult to say what
each implement was formerly used for, and still more difficult
to do this from any individual portion. It is well to form one's
opinion from the accounts given of savages living at the
present day. Unfortunately these accounts are often not
sufficiently exact, or they go into too few details, or they do
not give such particulars as would be of interest to archæology
and the study of prehistoric man.

On some specimens of this second kind of spear-heads we
find certain ornamentations, which, together with what has
been already mentioned, indicate an essential advance, and
probably also a later origin. This ornamentation, which is
again found on the upper side, consists generally of two lines
running parallel, and extends from the furrow to the chisel-
formed plane. From each of these lines there are drawn, sometimes inwards and sometimes outwards, a number of smaller
diagonal and parallel lines as an ornament of the whole thing.

Amongst the group of lance-heads there are six specimens
which have but very little of the point left, and which have
entirely lost the opposite end (Plate VII. fig. 30). These
fragments are all slightly curved, and very singularly are all
ornamented in the same way. The ornamentation consists of
two parallel lines on the upper side about ·118 inch dec.
apart, the space between which is filled up with diagonal
parallel strokes. Many of the implements of the Indians
of the present day are strikingly similar in their ornamentation, and we are involuntarily led to ask: Is it possible
that these Indians and the cave-dwellers of the Kesslerloch
can be the descendants of one and the same race, and
has this art of drawing been transmitted from this race
to all the others, or is it an original invention of different
races? I most certainly am of opinion that the human race
sprang not from one original pair only, but from several, and
consequently I believe that a transmission of this art is in
this particular case not to be thought of. This may sound
somewhat paradoxical, and against the facts which have been
already mentioned. But we know of similar cases even in
historic times; as, for instance, the telescope was invented
at the same time by a Dutchman and an Italian. As a
general rule it may with safety be assumed that different
races of men are more or less compelled by similar circumstances and influences to make use of similar means. Our

Kesslerloch artist in the exercise of his art made use not only of lines, but of dots. Thus we find, for example, on the upper side of a broken lance-head about 10¼ inches long the ornaments just mentioned, and on the right side two lines twisted in a serpentine form, parallel, and formed of a succession of dots close to each other. Under these two sinuous lines there is a horizontal line, also consisting of a series of dots. The regularity of this ornamentation indicates an expert eye and a considerable quickness of hand.

There is another kind of lance-head, thick in the middle and tapering to a point at both ends (Plate VI. fig. 27). This is the only specimen of the kind found in the cave; but in the French caves, and in the neighbouring cave of Freudenthal a considerable number have been found. This implement was probably hafted by the shorter point being fixed into the porous part of an antler.

The *harpoons* belong to the third group of bone implements; compared with other projectiles, their number is very small. Altogether only eight specimens have been found, and they differ very much from one another, both in their workmanship and also in their state of preservation. They may be divided into two classes: those which have the barbs on one side, and those which have them on both. Of the first class we have three, and of the latter five specimens. The first, with the exception of a few barbs, have been preserved entire. One is nearly 6¼ inches long, and in the middle it is nearly a quarter of an inch thick. Both ends terminate in sharp points, one of which is somewhat round, while the other is spread out on the left side so as to form a sharp edge, in which there are several notches or incisions (Plate VII. fig. 35). Both points are about 1¹⁄₁₀ inch long. The barbs, five in number, two of which are broken off, become larger, and are wider asunder in the hinder part; they have been made with wonderful ingenuity and neatness; their points are as fine as those of our finest steel needles; they are not made to stand at right angles, but run parallel with the shaft.[1] This specimen is the best preserved, and also the best executed of any that have been found. The other two specimens of this class (Plate VIII. fig. 48) are very similar to the one just described, with this difference, that one of them at the hinder end is expanded like a ball, similar to those found in Périgord. The second class of harpoons, which are barbed

[1] Harpoons with barbs very similar in form, but on both sides of the weapon, are drawn in *Reliquiæ Aquitanicæ.* B. Plate I. fig. 1, and B. Plate XIV. fig. 1.

on both sides, undoubtedly indicate a greater amount of skill, and must have been more effective than the others, as they would remain fixed in the body of the animal for a longer time. Unfortunately not a single specimen is entirely perfect; they are all defective either at one end or the other (Plate VIII. fig. 47). The specimen of this kind which is in the best state of preservation is drawn, Plate VI. fig. 25: it is about six inches long, and about a third of an inch wide; it has seven whole barbs, and one broken off; all of them are made just like those of the harpoon first described; they stand opposite to one another, and are oblique, so as to enter more easily into the body of the animal struck. Both on the upper and under sides there is an ornamentation consisting of several double straight lines placed obliquely. The four remaining examples of this class have no essential peculiarities, one of them is drawn (Plate XIV. fig. 94). It is evident that the harpoons were stuck in a shaft, and as, with a single exception, the part going into the shaft was pointed. I imagine that it was stuck into the porous part of the horn, into which the point could be forced with little trouble. But in order to prevent the harpoon when struck rebounding and pressing too deeply into the horn, and thus weakening the effect of the stroke, these points were very wisely made either ball-shaped or spade-shaped. It appears to me that the harpoons found in the Kesslerloch were exclusively used as missiles for birds. The especial use of the barbs was to make the harpoon fast in the flesh of the game, so that when it got up into the air, and was escaping, it was prevented from doing so by a string fastened to the barb, and so was captured. The cave-dwellers at St. Madeleine perforated the harpoon in order to fasten the string more easily to it, and the same thing is done by the Esquimaux of the present day. They use harpoons more especially in hunting seals.

It cannot be decided with any certainty how the arrows, lances, and harpoons were darted against the prey; probably these cave-dwellers, like the savages of the present day, used bows made of wood and catgut. The modern 'Snake-Indians,' with similar weapons, strike their prey thirty or forty paces off with great precision, and with such force that they can send their arrows through the body of a horse or a buffalo. If bows like these had been used by the inhabitants of the Kesslerloch it is evident that they could not have lasted till now, as the wood in the course of thousands of years must have decayed.

The fourth group consists of what are called *scrapers*, and

belong more to every-day life than those last mentioned. They
are very simple implements, being only rounded at one end, and
somewhat sharpened on both sides. Every specimen shows
more or less distinctly on its edges the cuts which were made
when the horn was divided in pieces, and in every case the
under side shows the porous part of the horn. These implements
vary greatly in size; they are from 8 to $2\frac{1}{2}$ inches long, and
from nearly $4\frac{1}{2}$ to $\frac{3}{4}$ of an inch broad. It does not appear to
me that these little scrapers had been broken off at the end
which is not rounded, for amongst the sixteen specimens found
there are six which are all of the same length, and fit the hand
of an ordinary man very well, if the thumb is placed at the
thicker end (Plate VII. fig. 31). Doubtless these implements
were used for skinning animals, as they were very well adapted
to this purpose. Similar implements have been found in tolerable abundance in the Belgian caves. One specimen deserves
especial mention on account of its ornamentation. It is well
formed, and equally broad throughout its whole length, and one
end is neatly rounded; the other end unfortunately is broken
off. Both above and below it is so well polished as, when
touched, almost to feel like glass. On the upper side, near
both edges, there is a row of raised points, which when
examined by a magnifying glass, have all more or less
decidedly a rhombic form. Several incised lines are drawn
lengthwise on the very edges. It is singular that on the under
side there are some incised lines, which may possibly indicate
that at an earlier period this implement may have been used for
some different purpose (Plate VII. fig. 29). It is very singular
that we found but one implement made from a medullary
bone; it is sharpened on both the borders or edges very neatly,
and towards the end it is pointed almost like a dagger.
No better instrument could have been devised for skinning.
Three specimens of worked ribs found here were probably used
for the same purpose; they are finely polished on both
sides, and are rounded and sharpened like the scrapers at one
end; one specimen is rounded and sharpened at both ends
(Plate VI. fig. 28).

The specimens which are called *awls*, or boring implements,
are just as simple in their construction (Plate VI. fig. 24,
and Plate XIV. fig. 89). They are made of some convenient fragment of bone, neatly sharpened at one end. They evidently were
used just as the awls of saddlers and shoemakers are in our

own days, to make holes in the skins, so that a thread might be
drawn through them, and thus several single skins might be
sewn together. No great number of these tools were found;
only three specimens in all. Still a larger number must have
been made, and as they were easily broken when used, the
broken ones were thrown aside as useless, and would be replaced
with very little trouble.

As the operation of drawing thread by the hand alone
through the holes made by the awls must have required both
time and patience, it was natural that needles should have been
called in to aid. They were found in tolerable abundance;
but of the twelve specimens which came to light only four were
perfect. They vary in length from more than 2½ inches to
1¾ inch. At one end they run to a fine point, while the other
end is somewhat flattened and rounded off, and is furnished
with a neat round eye to receive the thread. The whole needle is
well polished, and in general is as well finished as a steel needle.
The eye is so fine and well made that it seems almost impossible
to understand how it should have been made with such in-
sufficient tools (Plate IX. figs. 64 and 65), and yet this was
actually the case. I have convinced myself of it by repeated
trials. I set to work to make one of these eyes by boring from
both sides a needle which had been scraped into shape, and
continued this manipulation till the perforations on both sides
met. If bored only on one side there was a very great risk of
breaking the flint point, besides which the diameter of the eye
would have been much too large for the part to be perforated.
When the needles of the Kesslerloch are closely examined it
may be seen very distinctly that the cave-dwellers must have
used the same method of boring their needles, for the eye is
always the smallest in the middle. I have tried to sew with
these bone needles, and was perfectly successful with common
linen. I did not venture to try it with thicker material, and it
seems to me impossible to do so without breaking the needles to
pieces. Two needles found at a later period in the rubbish from
the cave are very singular. They are added here by the kindness of Dr. Ferdinand Keller (Plate IX. figs. 62 and 63). In
these two specimens the eye is oval, just like that of our larger
sewing or darning needles.[1] There can be no doubt but that
they were used for thicker thread. They must have been much

[1] Bone needles with oval or oblong eyes have been found in the cave at Lourdes.
'The cave at Lourdes (Hautes-Pyrénées), containing many remains of the reindeer, has

more difficult to make than those previously described. One of the specimens was never finished, for the part above the eye has not been rounded off. From this it would seem that the eye was probably made first, before the needle was sharpened at one end and rounded at the other, so as to save time in case the needle should break in the process of boring, which doubtless was frequently the case. To the best of my knowledge these two specimens are the only bone needles with oval eyes which were found in the cave. They probably form a transition to the bronze needles, which, if I mistake not, have chiefly oval eyes. [1] These two needles are nearly $2\frac{1}{4}$ inches in length; they are thicker than the others, and look more bulky. All the needles were found in the upper relic-bed, and they evidently indicate a more advanced degree of civilisation amongst these prehistoric men, for doubtless with these needles they made the reindeer skins into clothes.

It can hardly be said what was the use of the implement drawn, Plate XIII. fig. 73. I do not remember ever having seen drawings of similar specimens in bone. A second, almost exactly similar, was also found (Plate XIV. fig. 88). The ornamentation of one of them is good, and indicates a considerable degree of artistic talent.

Just as little can be said as to the use of the bone implement drawn, Plate VIII. fig. 46. It is thin, and nearly three inches long, and is brought to a point at both ends. [2] In the cave at St. Madeleine similar implements were found in tolerable abundance, and M. Lartet believed that they may have been used as fish-hooks, when stuck diagonally into another piece of bone. This

---

furnished only two coarse needles having an oblong head and eye, not pierced by boring, but rather by cutting with a sharp instrument. One of these needles has been illustrated by M. Alphonse Milne-Edwards in his note on the works of man found in this cave at Lourdes.'—*Ann. des Sc. Nat.* 4me. Série Zool. vol. xvii. 1862, p. 243, Pl. VI. fig. 3.

This quotation is from the *Reliq. Aquit.* p. 140. The whole of the treatise is most interesting; it is by the late M. Lartet, and commences page 127. The title is 'On the employment of sewing-needles in ancient times.'

[1] See Keller's *Lake-Dwellings*, 1866 (English translation), Plate XXXVI. figs. 8, 12, 13, 14, 15, and 18. The whole of these are of bronze, and were obtained at the lake-dwelling of Nidau.

[2] It is perhaps not wise to assume that similar implements have always similar uses; but there is most certainly a striking resemblance between this implement and that figured from Wangen by Dr. Keller in his *Lake-Dwellings* (English translation), Plate XIV. fig. 23. Dr. Keller says that these implements occurred at Wangen rather plentifully, and he seems to consider them as fishing implements. He adds, however, that they are now in use on the Untersee for catching ducks.

idea cannot well hold good here, for amongst all the animal remains found in the cave not a single fish-bone has been noticed. Possibly they may have been intended for a kind of awl or piercer, one end having been fastened into a handle.

There is also another singular implement drawn, Plate VIII. fig. 44, and between five and six inches long. At one end it is ground down on one side like a chisel, and towards the end of this plane there is a hollow of a somewhat curved form and about a third of an inch deep; probably this may have been for the insertion of a flint. At the other end there is a cavity rather more than a quarter of an inch deep; but whether this was due to chance or intention cannot be decided. Though this implement very much resembles the agricultural tools of the lake-dwellers, yet I very much doubt whether it was used for a similar purpose. May it not have been a tool for extracting the edible roots from the ground?

It is difficult to say to what purpose the implements drawn, Plate XIV. figs. 91 and 92, and Plate IX. fig. 55, were applied.

The specimen drawn, Plate VII. fig. 42, probably represents a dagger with a carved handle, the point of which, however, is unfortunately broken off. Even the ornamentation has suffered considerably by use in the course of time.

The two implements drawn, Plate VIII. fig. 45, and Plate XIV. fig. 93, are interesting. As far as I know similar implements have not been found in any other cave.[1] On the larger specimen there are very peculiar engravings, hardly visible to the naked eye; parallel lines on a raised ground alternate with similar lines in hollows.

Together with these different implements there were found a considerable number of objects used for ornaments. Every one of these is perforated; the holes are small, hardly a line in diameter, and very neatly worked. We may first mention five teeth perforated at the root, three of them are incisors of a horse, another is the canine of a small carnivorous animal, and the fifth is indeterminable. This last (Plate XIII. fig. 78) is in fact cut off at its root, and at the other end it has been rubbed down, so that its upper surface has a black and yellow striped appearance. No doubt this ornament was highly prized in those days. Most probably all these perforated teeth (Plate

[1] My friend, Mr. H. Woodward of the British Museum informs me that similar implements have been found at Bruniquel. He tells me that they are considered as being harpoons or javelin-tops.

IX. figs. 53 and 56), and in general all the smaller objects which had holes in them, were used as earrings or neck-ornaments. A thread was passed through the holes, and then tied in a knot to the ear or round the neck. At all events they are very simple ornaments, and these objects have been found in nearly every cave which has ever been inhabited by man.

Two flat perforated slabs of bone were also found (Plate XIII. figs. 76 and 77) ; at one time the form had been round. One of them has on both sides branch-like ornaments in a radiating form.

The earring drawn, Plate IX. fig. 57, looks more like those of the present day. It is of bone, and finely polished. Probably also in this case the perforation may have been made first, and then afterwards the curved form immediately adjacent was given to the ornament.

The ornaments of 'brown coal' are unique of their kind. Where the cave-dwellers got the material it is impossible to say. It is well known that the Jurassic limestone, and the marly slate, and the sandy limestone immediately overlying the uppermost beds of the lower white Jura, occasionally contain what are sometimes called 'pockets' of coal. It may therefore be quite possible that these coal ornaments may have come from places like those in the immediate neighbourhood. At the present day small pieces of coal are to be found at Schienenerberg, near Ramsen, and we may, perhaps, conclude that the pieces of coal found in the cave were got from this locality. The ornaments were worked exactly in the same manner as those of bone, as may be seen, Plate IX. fig. 58. All these ornaments are in good preservation, and have been worked with surprising neatness; they also have been used as earrings or neck-ornaments. One of them (Plate XIV. fig. 82) is circular and about an inch and a quarter in diameter; it is perforated in the centre, and it becomes somewhat thinner towards the circumference. On both sides there are a number of scratches made with flint, the last traces of the grinding or polishing.[1]

Two amulets are very interesting ; they are nearly alike in form, and certainly would not be despised by the ladies of the present day, any more than by those of the prehistoric age. One of these earrings (Plate XIV. fig. 83) has no further ornamentation, except being polished on the convex side, and having in the middle of it a kind of oval escutcheon. The other,

[1] This specimen looks very much like a spindle-whorl of somewhat later date.

on the contrary (Plate XIV. fig. 85), has on its upper convex side a number of neatly-engraved lines formed of very small dots. The work is charming, and so well executed that it would require considerable expertness to imitate it.

An earring executed with similar dexterity is drawn, Plate IX. fig. 59. Instead of being ball-shaped, it takes a longer form, and gradually narrows from top to bottom. On the front, or narrower side, there are two lines gradually approximating, and, like the specimen last mentioned, formed of dots.

An amulet not quite finished and not perforated was also found (Plate IX. fig. 60), of a flattish cylindrical form. A specimen exactly similar was subsequently found in the rubbish from the cave. Some other similar earrings are drawn, Plate IX. figs. 52, 61, and Plate XIII. figs. 75, 79, and 81.

The ornaments are not all made of bone and coal; some are of stone or are petrifactions (ammonites), as may be seen from the specimens drawn, Plate VII. figs. 37, 38, 39, and 40; and Plate XIV. fig. 95.[1]

Even shells were used as ornaments. Several were found in our cave: one of them has been ground down near the hinge, till a hole was made; another has been twice perforated artificially (Plate XIV. fig. 84), together with these worked shells; others were found in their natural state. Professor Karl Meyer of Zürich has had the kindness to determine them for me, and I return to him many thanks for so doing. The following species are mentioned by him:—*Pectunculus glycymeris, Ostrea cucullata, Pectunculus Fichteli,* and *Cerithium margarit-aceum.* He considers the two first named to be recent shells; but the others he believes to be fossil.

The last group of implements to be mentioned are the larger portions of bone with holes bored through them. Most strangely, not a single specimen is entirely perfect. All are broken off at one end, and on this account have been thrown aside as useless into the rubbish of the cave. No less than twenty-three bones of this kind were found with a single perforation, and four of the same sort with two holes. They vary in length from more than 12 inches to $3\frac{1}{2}$ inches, and in thickness from $\frac{1}{12}$ to $\frac{1}{7}$ of

---

[1] Amongst the things found in the rubbish-heap after the excavation was finished was a very beautiful fossil tooth of the *Ceratodus,* apparently the *C. serratus* of Agassiz, from the 'Muschelkalk,' two inches across. It must have been brought to the cave as a curiosity; but its presence there shows that the inhabitants roamed to a considerable distance, or they must have had a barter trade with the neighbouring races, for the nearest point where the 'muschelkalk' occurs is said to be three leagues distant from the cave.

an inch. The diameter of the holes varies also very much, viz., from $\frac{1}{4}$ inch to very nearly an inch. The smallest of these bones are ornamented on the upper surface. MM. Vogt and Dupont consider these perforated implements as 'commando-staffs' (or bâtons of office) because the Indians have similar distinctions for their chiefs. (Plate V. fig. 17, and Plate VIII. fig. 43.[1])

It can hardly be divined for what purpose the doubly perforated bones were used, which are drawn, Plate VII. fig. 41, and Plate IX. fig. 49.

Another bone has been perforated in a very peculiar manner. The hole has been made in such a direction through the bone, that a staff or pole forced through it would not be at right angles to the bone, but nearly in the same line with it. The bone is hollowed out in some measure on both sides near the hole, and everything seems to indicate that a staff has been driven through the opening. According to our ideas this implement bears some likeness to a kind of spade or spud, certainly of a very primitive description, but which the cave-dwellers may possibly have used to get edible roots from the ground.

Having finished these remarks as to the different classes of implements, I will now pass on to the more valuable art productions of that age; namely, the engraved drawings on the bones, some of which far surpass in execution all that has been previously known of the kind. [2] A drawing, somewhat defectively executed, was found on the broken brow-antler of a reindeer (Plate XI. fig. 67). The drawing is of some animal, but unfortunately only the hinder part has been preserved. The outline is sharply marked, more especially the line of the back. The small curled tail and the clumsy build of the body lead to the idea that it was intended for a pig, but there are no further indications which may help us to decide as to this figure. On

---

[1] Several of these 'bâtons' are drawn in *Reliq. Aquit.*. Plates B. XV. and XVI. On that drawn, Plate VIII. fig. 43, and also on several other bone implements, there are certain peculiar notches or marks which are very similar to those so well described by Professor T. Rupert Jones in the 13th and 14th numbers of *Reliquiæ Aquitanicæ*, and which he considers to be 'marks, indicative of ownership, tallying, and gambling.' But may they not have been mere ornaments? I have in my possession a pipe, of which the bowl is clay, but the stem and mouthpiece are of wood and horn; it was given to me by Mr. Geo. H. Kinahan, of the Irish Geological Survey, and was manufactured (or at least the stem of it was) by some of the natives on the west coast of Ireland. This stem is ornamented in a very 'prehistoric' (!) way, and I am told that the natives there have no better tool for the ornamentation than a common kitchen knife.

[2] This, as already mentioned, is doubted by Mr. Franks.

the back of this piece of horn, when closely examined, may be seen a number of lines, which probably were intended for a drawing, but which have entirely lost their character by decay.

On the 'beam' of a reindeer horn (Plate XIII. fig. 72) about seven inches long, from which the brow-antler has been taken in the usual manner, there is the drawing of a head, the outlines of which are partly in good preservation. The upper part of the head is no longer to be seen, as the horn has decayed near the frontlet. In the place of the eye there is a small circular hollow; below the neck there are a number of striæ, which doubtless are meant to represent hairs. The drawing in itself is extremely coarse and hastily executed, but at the same time an unskilled artist would hardly be in a condition to engrave such a figure on horn. The mane under the neck indicates that it was a reindeer. The whole horn is finely scraped, and in front has been broken off and rounded. Probably this instrument may have been used as a dagger.

We find some drawings even upon coal. A piece of this substance, about 1½ inch long and 2¼ inches wide, has on either side of it a head engraved, one of which is less definite and less neat than the other. In both the figures (Plate XV. figs. 96 and 97 [1]) we again find a long mane under the neck, and also

[1] In Professor Rütimeyer's last work, to which we have so often referred, he considers (in the Appendix) that the figures of these horses' heads, given by Mr. Merk, are

not quite accurate, and he therefore gives in his little volume woodcuts, which he says are much more correct; these woodcuts we have ventured on copying by the new 'Bank' process.

on the nape, so that we may conclude that the artist intended to draw the heads of two horses. The eye in one of the figures is very much more artistically drawn than in the other. We find not merely a circular depression, but we can even distinguish the eyelid and the eyeball—here, however, the ears are wanting; while in the other drawing there is an erect ear terminating in a point. Even the hairs about the nostrils are given. On the whole both drawings have been executed with much skill.

We have one more drawing on a piece of coal, but the meaning of it cannot be deciphered. The piece is elliptic in form; the greater diameter is about 2¼ inches, and the smaller diameter 2 inches. The lines are clearly marked, but run together quite confusedly (Plate IX. fig. 50).

But drawings of entire animals from the Thayngen cave lie before us. On a piece of the main branch of a reindeer's horn about 12½ inches long and 1½ thick, three figures of animals have been engraved, but unfortunately have suffered from the decay of the horn (Plate X. figs. 66A and B). The upper and broader end of this piece of horn is perforated, and the hole is nearly an inch in diameter. One of the figures has its head immediately above the hole; the other two stretch their heads in an opposite direction. The figure mentioned first is best preserved, and though the execution is decidedly defective, yet it is unquestionably better than the other two. The head is drawn very much more neatly, and the eye more especially shows a marked improvement, if we may venture so to speak, for it is hard to say whether these drawings are the productions of the same artist. If the same person had made both, most decidedly the difference between the two first and this one is so great that we may almost conclude that a great many intermediate attempts of this kind must have been lost outside of the cave, or have been destroyed, and that only the smallest portion has come down to us. But if, on the other hand, the art of drawing was a common qualification of our cave-dwellers, then the drawing of the animal under consideration must have been executed by a very advanced artist. The long erect mane on the nape of the neck, the beard under the throat, and the tail (of which part only remains) indicate very decidedly the wild horse of that day. The hind legs are very stiff, while the front legs are drawn much more naturally. The whole weight of the fore part of the body rests for a moment on the left fore-leg, which in fact bends a trifle, while the right leg has made

a step forwards and is just about to touch the ground again.
This position of the animal gives life to the whole, and indicates a considerable amount of observation. The form of the
foot is singular, as it resembles that of the reindeer more
than that of the horse. The two other figures, which are also
drawn stepping forwards, can only be known to represent
animals by coarse and broken outlines. They may probably
have been intended for reindeer; at all events the short and
stumpy tails, which are rather indistinctly shown, incline me
to this opinion. In both figures the head is entirely wanting.

The real artistic production of that age is the drawing
of the reindeer, which is already well known, and which has
been previously published (Plate XII. fig. 71). It is engraved
on the 'beam' of a reindeer's horn, more than 7 inches long
and $1\frac{1}{6}$ of an inch wide. This 'beam' is not in good preservation, in fact it is broken off at both ends, and yet at one
end it was perforated, as half of the hole shows. The hole
was nearly an inch in diameter. This fragment, with the
perforation, was at first taken to be a complete portion of a
reindeer's antler. The whole horn is finely scraped. Round
the perforation, and on the under side, there are a great
many scratched lines, and also one tolerably deep and broad
furrow, which is not to be found in the other similar instru-
ments. Evidently the only use of this furrow was, as before
mentioned, to divide the antler into pieces in the usual way;
but during this operation it occurred to the workman to
employ the antler for another purpose, and he drew upon it
the likeness of an animal which was continually before his
eyes. This figure, even at a superficial glance, may at once
be recognised as a reindeer. The horn with its brow-antler
and side 'tynes,' the mane on the front part of the neck, and
the short, stumpy tail, are unmistakeable characteristics of
this animal. The lines of the back and the hind-legs are
the most strongly marked. The fore-legs, the line of the
body, the head, and the antlers are less decided. The neck
and the breast are very indefinite, so that the connection
between the head and the rump is but faintly made out.
The outlines of the hoofs also have lost their distinctive
character. The three lines under the body are very singular.
Though an artist of the present day might have much to find
fault with in this figure, yet it would be impossible not to
wonder how man, thousands of years ago, should have made
such progress in art as to produce these drawings with such

E

primitive tools; in fact, it seems almost impossible. But
we have similar phenomena in other quarters: there are
in fact whole peoples, as well as individual men, who·in certain departments of human science and knowledge have
made great advances above others for centuries, nay even for
thousands of years. Are not the Greeks by their former
skill in the plastic arts and poetry a pattern to us even at
the present day? Just so we may consider that our prehistoric artist has in his department so far surpassed his
compeers, the cave-dwellers of Périgord and Dordogne, that
neither their ancestors nor their successors were able to exe-
cute such art productions with flints. The reindeer as it steps
forward grazing brings in fact before the imagination a
peaceful image; so that our artist can hardly have been a
savage rough hunter, whose great enjoyment was in wild
romantic hunting scenes. The general bearing of this reindeer stepping forwards, and that of the horse already mentioned,
may possibly indicate that the idea of thus representing
animals sprang from the same brain, and perhaps we may have
to attribute to one artist the drawings of the horse and the
reindeer.

The same claim to artistic skill may be made by the
drawings of the horse represented, Plate XII. fig. 70, and Plate
XI. fig. 68. Here the figure of a horse is drawn on a piece
of reindeer-horn about 9 inches long and 1½ of an inch thick,
which is perforated at the end. The fine outlines and shading
work appear as sharply cut as if they had been engraved but
yesterday. Unfortunately the horn was carelessly broken in
two when dug up; by good chance, however, the drawing has
suffered but little. As the artist, on account of the limited
space, could not quite place the drawing on the upper side of
the horn, the hoofs have come on the under side, and thus
lose somewhat of their significance. The parallel striæ on the
neck, the breast, the back, the body, and the hinder part
of the animal undoubtedly represent hairs. The well-formed
head—rather long, with small ears—the upright mane, the
graceful, well-formed body, the elegant and lightly-formed
feet, and especially the remarkably thin tail, reaching nearly
to the ground, represent without doubt a young well-bred
animal. And here again the horse is drawn stepping forwards. One might have been tempted to have considered this
drawing as the representation of a cross between the horse
and the ass, but the horses' bones found in the cave show that

the animal of those days in no way differed from our tame horses of the present age.[1]

Our cave has yielded us not only drawings, but actual sculptures. Unfortunately they have not come down to us perfect, but still they are such as to interest us extremely. Plate XI. fig. 69, is the head of an animal with part of the neck.[2] The two ears, the eyes, and the greater part of the horns are quite perfect; the muzzle and the nose are altogether wanting. Very strangely, one ear stands higher than the other, probably arising from the fact that the artist has made one horn broader than the other, so that he is decidedly in fault. And as the two eyes also are not at the same height, it seems as if the artist had worked on one side without referring to what he had already done on the other. The eye itself is very neatly carved. It projects a little, and has a decided eyelid. The horns, which are singularly broad at the base, rise quite in front and from the top of the forehead, so that this is here reduced to merely a very narrow hollow; they curve downwards at the side of the head, and seem to bend more forwards than backwards. The neck is ornamented with many lines, formed of a series of small strokes. It is difficult to say what animal this sculpture is intended to represent. Our list of horned animals comprises five species, viz.: the ibex, the chamois, the wisent, the urus, and this animal now under consideration—either the domestic ox or the musk-sheep. The sculpture most completely excludes the three first,

[1] These drawings have been shown to a veterinary surgeon, and also to a friend who has seen a great deal of cavalry service, and both of them think that the *form* of the animal is more like that of the zebra than of the horse.

[2] Annexed are the copies of the figures given of this curious sculpture in Professor

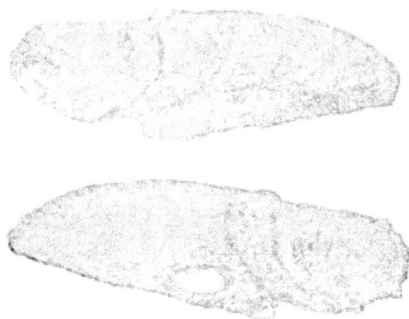

Rütimeyer's book above referred to. Professor Rütimeyer did not consider the litho-graphs perfectly accurate.

and also the domestic cattle, so that we have to decide between
the urus and the musk-sheep; but as the form of the horns
actually will not agree with those of the urus, my idea is
that the only animal which this sculpture can represent is
the musk-sheep. At the lower or hinder part of this specimen
the porous part of the horn is seen, so that we may conclude
that this sculpture is merely a fragment, and that the whole
may probably once have been the handle of a dagger. This
view is borne out by the fact that in the cave at Laugerie
Basse an actual dagger was found, with the handle formed of
a carved animal's head. The size of the specimen now under
consideration is as follows: it is about $2\frac{1}{3}$ inches long, and
$\frac{7}{10}$ of an inch thick in one direction and $\frac{5}{10}$ in another. A
second sculpture (Plate IX. fig. 51) was also found. It is a
regularly carved head, but the right ear is lost. The specimen
is rather more than $\frac{8}{10}$ of an inch long. The head runs quite
to a point; the nose is long, and not arched; the eyes are at
the same height, and the eyelid and eyeball are discernible.
The forehead is also clearly marked, with the protuberance over
each eye. The ear is rather narrow, and stands upright. The
specimen decidedly represents the head of a horse, and this
appears quite clearly when it is looked at either before or
behind. The whole head is ornamented with many parallel
striæ. The snout of this figure evidently rested upon a second
head, for on the under side, both to the right and the left,
two very definite ears may be seen, formed exactly like that
of the specimen under consideration. The musk-sheep's head
before described gives the idea more of a carving in relief, but in
the present case we have to do with a head regularly sculptured,
and consequently more with the plastic representation of an
animal's head.

As before mentioned, not a trace of pottery was found in
the relic-bed. But we most certainly met with a fragment of
pottery in the upper bed of rubbish which rested upon the
relic-bed (Plate XIII. fig. 80). This fragment has eight
circular depressions in it, and the material, like that of the
pottery in many of the Swiss lake-dwellings, is black mixed
with white bits of gravel. In this fragment we have therefore
an irrefragable proof that the cave was known, but not inhabited, in the time of the lake-dwellings.

No bronze or iron implements were found in the Kesslerloch,
but several pieces of iron ore, as, for instance, iron pyrites and
red oxide of iron. These were probably brought home by the

cave-dwellers, not on account of the iron they contained, but for their colour, and perhaps for their rarity. Amongst these pieces one was found of a remarkably oval form, which evidently had been a rolled pebble from the river-bed, and by some chance had got into the fire, for there are a great many melted granules on the surface. Probably in after years it was by some such chance as this that iron was discovered.

Now that I have finished my description of the different specimens which have been worked by the hand of man, I will add, for reference, the following tabular list of the specimens found in the Kesslerloch:

1. Flint-flakes . . 12 000
2. Pebbles used as hammers . . 200
3. Fragments of worked reindeer-horn 100
4. Arrow-heads . . 54
5. Lance-heads without a tarrow 95
6. Lance-heads with a farrow 49
7. Curved lance-heads . 6
8. Harpoons . 8
9. Scrapers . 16
10. Worked ribs 3
11. Awls or piercers 3
12. Needles . . 12
13. Undetermined implements . 7
14. Pieces of coal, worked and unworked . 60
15. Bone carrings . . . 3
16. Perforated teeth . . 5
17. Coal earrings . . . 14
18. Worked shells and ammonites 1
19. Bones with one perforation 23
20. Bones with two perforations 1
21. Drawings of heads . . 3
22. Drawings of animals . . 8
23. Sculptures . . 2

If we compare the objects found here with those from the lake-dwellings of East and Central Switzerland, it must be confessed that in many respects there is great similarity between the two. Flint-flakes, awls, arrow- and lance-heads, needles, and perforated bones are found in both these human settlements. On the other hand, we find at the Kesslerloch no pottery, stone celts, perforated stone implements, fish-hooks, or polishing stones; no spindle-whorls,[1] cloth, worked flints, nor a single implement destined for agriculture. But again, on the other hand, the specimens from the lake-dwellings do not exhibit any drawing or sculpture, or ornamented bone implements, or any

[1] Unless the object drawn, Plate XIV. fig. 12, should have been one, which, however, is very doubtful.

ornaments of this nature. For these reasons the lake-dwellings are not to be placed in the same age as the human abodes found in caves. Consequently the era of the cave-dwellers forms the background in the picture of prehistoric man.

If we compare with our specimens those found in the inhabited caves of other districts, as, for instance, the cave at Salève, at the opposite end of Switzerland, or the Belgian, French, and Swabian caves, we again find so much agreement, so much that is similar, that involuntarily we are led to the conclusion that the human race at that time in Switzerland, in Belgium, France, Swabia, England, and other countries had arrived pretty much at the same degree of civilisation, had felt the same needs, and in general existed under the same conditions of life. Probably the objects found in our cave have the greatest similarity with those of the various caves of Périgord and Dordogne. A comparison with the specimens from the French and Belgian caves has convinced me that our cave at Kesslerloch, though not very large, has yielded results which indisputably, both in quantity and quality, are of the richest and most interesting nature. Only two similar 'finds' have been discovered of late years in Switzerland, viz. that at Veyrier on the Salève, and that at Villeneuve. The cave at Veyrier was examined in 1834, and that at Villeneuve in 1870. Still more lately, in the neighbouring district of Freudenthal, a cave has been excavated by Professor Karsten and President Joos of Schaffhausen, and the specimens found here, though not numerous, are of a highly interesting character. In another cave at Dachsenbuel, about a quarter of an hour's walk west of Herblingen, Dr. Von Mandach found a number of portions of human skeletons, some flint-flakes, a neck-ornament, and a bone implement, together with a good deal of pottery. It is therefore certain that the whole of this district was in early times the theatre of human activity.

If we now return to the consideration of the cave-dwellers of our Kesslerloch, and picture to our imagination for a few moments their life and occupation, a very peculiar idea strikes our mind. The life led by man in those far distant times was in many respects very little better than that of the beasts which he hunted. It was in the fullest sense of the words a constant struggle for existence. It is very evident that caves were selected by him as places of refuge and habitation. He did not at that time understand how to build huts to protect him, nor had he, in fact, the tools necessary for this purpose. A cave, therefore, with its sheltering roof, offered to him a welcome abode to secure him from storm, rain, and cold ; in fact from all the hurtful influences of nature.   It was very natural that the caves of the Jura formation should be inhabited before others ; for no other formation has so many caves as this, besides which, in general, flint (or chert) was found in it, which was then so necessary to human existence.   Caves situated in the Nagelfluh have also been inhabited ; as, for instance, that at Villeneuve. There are many different views as to the mode in which such caves were formed.   Dupont thinks that they are caused by hot springs ; other scientific men say from volcanic upheavals and depressions ; and others, again, that they have been caused by the action of water.   Naturally all caves cannot have been formed by the same causes.[1]   No one who has seen the Kesserloch can doubt that it was caused purely by the action of water.   No doubt it was at one time filled with loam or clay (Lehm), like other caves which are brought to light occasionally, when the Jura limestone is blasted.   But when the valleys were formed this loam was then laid bare, and was washed away by the running water.   There can be no doubt that the Kesserloch was a favourite place of abode, for hardly any cave is better situated or more convenient inside.   The cave at Freudenthal compared with this is a mere nook.

The animals amongst which our cave-men lived were very varied, as the fauna shows.   The most harmless animals, such as the marmot, the hare, the chamois, &c., and the wildest beasts, such as the bear, the lion, the wolf, &c., were merely the contemporaries of man ; for no one will contend that he lived *in the cave together* with the animals whose bones we found scattered about there ; more especially if it be considered that the flint knives and bone implements made by man were found, together with the bones, in every bed and in all parts of the cave.   The companionship of man was in fact by no means a peaceful one.   Most of the animals were very superior to him both in size and strength, and this considerably aggravated his struggle for existence.   He had therefore to endeavour to make up by his intelligence that which was defective in his bodily powers.   He was obliged to be the master of these wild beasts if he wished to attain that security of existence which is naturally so earnestly longed for by man and beast.   In this struggle man

---

[1] The reader will find a great deal of information as to the formation of caves in the very interesting work, lately published, called *Cave Hunting*, by W. Boyd-Dawkins, F.R.S., &c.

remained the victor, as may be deduced from the fact that he used the flesh of these animals for food. But how he managed to overpower these giants is somewhat of a mystery to us. One thing is certain, that the cave-dweller of the Kesslerloch was exclusively a hunter, as has been often pointed out, and doubtless he secured his prey in a clever and dexterous way, like the Indians of the present day. He was equipped with arrow and lance, with dagger and bow. Hidden sometimes by a projecting rock, sometimes behind a bush, he watched the passing game; creeping on the ground, he got nearer to the grazing animal, and then either hurled or shot from his bow the deadly arrow into the body of his victim. His hunting-ground was not limited to the immediate neighbourhood, but extended to the Freudenthal and Hemmenthal, and even to Merishausen, as is indicated by the belemnites found in the cave. Our cave-dwellers seem therefore to have lived exclusively on the flesh of hunted animals which they roasted in the fire, and, as before mentioned, the marrow was one of their great luxuries. For in fact it is merely a supposition that they had tools for digging up eatable roots, though I cannot quite consider it as improbable. Another portion of their time was occupied in preparing flint implements and hunting gear. Probably the women had to prepare the food, to keep up the fire, to sew reindeer-skins together, and to make clothes of them. Every one had his work assigned to him, and no one durst venture to give himself up to idleness, if the whole family were to be victors in the struggle for existence. The love of adornment was very marked amongst our cave-dwellers. Their ornaments consisted chiefly of necklaces and earrings. Very probably, also, they painted their faces, as many savages do at the present day. I was led to this opinion by finding a slab about half a square foot in size, one side of which was finely ground down and regularly covered with red paint. Near this slab were found two pieces of soft ruddle, or red oxide, which evidently had been the colouring material. It is therefore a point tolerably well decided that the ancient inhabitants of Switzerland were people clothed in reindeer-skins, who made and adorned themselves with bone or coal earrings, carried weapons of bone, and painted their faces. How different was it then from the present day! Probably from the very desire of ornament arose the art of drawing, which, in the figures of the horse and the reindeer, was carried to a degree not before in existence. The opinion has indeed been expressed that these drawings very probably were not done by the inhabitants of the

Kesslerloch, but had come into their possession by exchange with some of the neighbouring races. But if we examine the drawings of all the other cave-dwellers, we shall find that they are by no means so well executed as those from the Kesslerloch. Whence could they have got drawings, which up to the present day have never been found so perfect as these? I have not, however, the slightest doubt that our cave-dwellers made exchanges with other cave-dwellers, as this may be proved by the shells found in the cave, some from the Mediterranean Sea, and others from the Vienna Basin.

I will take this opportunity of expressing an idea which may be worth more mature consideration. It seems to me, in fact, that it would further the objects of the Swiss Natural History Society, if, as has been done for geology, it were to grant a sum of money for the exploration of the Swiss caves. If this were done, we should soon arrive at new discoveries. We have decided proofs that our Troglodytes did not always remain at the same grade of civilisation during their residence in the Kesslerloch; for if we investigate the position where the different implements were found, it is very striking that the ornaments, needles, drawings, and sculptures occurred only in the uppermost parts of the relic-bed. Thus, for instance, the objects drawn, Plate XII. fig. 70, and Plate X. fig. 66, were found in the immediate neighbourhood of the pillar before mentioned, and hardly two inches and a half under the surface of the black relic-bed. If we had rigidly sorted all the specimens when found according to the beds in which they occurred, this result would have been shown in a more striking manner.

Probably no one doubts that the grade of civilisation of our cave-dwellers is to be placed in far distant times. Though this is not the place for a regular dissertation on these remote periods, yet it may not be out of place to bring forward some conjectures as to the age of this settlement. The first notice in history as to the inhabitants of Helvetia is found in Cæsar; but in his reports about Switzerland he does not say a word about the lake-dwellings, though the remains of two hundred or thereabouts of these settlements are known at the present day. It is hardly to be supposed that this man, of such a powerful and versatile genius, would not have known of these peculiar villages, built in the water, if they had existed in his time; and we may therefore with great probability assume that at that period, or about 60 B.C., the lake-dwellings had disappeared; for Cæsar says in his report that the Helvetians had settled in

twelve towns and four hundred villages. Probably many a
century may have elapsed between the occupation by the lake-
dwellers and their removal to villages on dry land. For how
long a time these lake-dwellings were inhabited we probably can
hardly decide. In general they are believed to have existed
from one to two thousand years; consequently, since the settlement of the first Swiss lake-dwelling at least from three to
four thousand years have elapsed. But if our cave-dwellers had
actually been contemporary with the lake-dwellers, they would
certainly have held mutual relations with them; for the Kesslerloch was distant from Stein, the nearest lake-dwelling, only
about a walk of two good hours, and the cave-dwellers would
certainly have learnt a very great deal from the more advanced
lake-dwellers. But the antiquities found in our cave do not
give the least trace or indication of the lake-dwellings. The
total absence of corn, of pottery, of bored stones, and of stone
celts, and also of domestic animals, and, I may add, the entirely
altered fauna, are very certain evidences that the age when
men lived in caves is to be sought for far beyond that of the
lake-dwellings. I have already stated that these changes of
fauna arise from changes of climatic relations. But climatic
changes do not occur in short periods; thousands of years are
required for them. The period during which our cave was
inhabited cannot certainly be accurately fixed; but, at all
events, many centuries must have passed before a relic-bed
nearly three feet in thickness could have been formed. Now
although by these reasonings as to the inhabitants of the Kesslerloch we cannot make out any determinate number of years
chronologically, yet they enable us to arrive at the conclusion
that many thousands of years must have elapsed since the first
settlement of man in the cave of Thayngen. One fact already
mentioned confirms this idea, and that is, that the lowest part
of the relic-bed is at least four feet under the present level of
the valley. But as the flat of the valley was formed by deposits
of gravel and sand, brought down to it, and as at one time in fact
it was under water, the Kesslerloch must have been inhabited
before this alluvial action. Very probably this alluvial deposit
in the valley arose from the Lake of Radolphzell, which at one
time had its outlet by Gottmadingen and Thayngen to Schaffhausen. Doubtless this outlet did not exist while the cave was
inhabited, although the lake was no great distance off. I explain
this fact by assuming that at that time the lake was probably held
back from running in that direction by the rubbish-heaps in

front. Besides this, it may be taken for granted that when the climate was cold there could be no very great quantity of water. The land was in fact in a great measure covered with glaciers. At length the great mass of waters, evidently caused by the altered warmth of the climate melting the ice, broke through the dam, and poured in two streams to Schaffhausen, on one side by Stein, and on the other by Thayngen. Certainly no gravel from this flood has been found inside the Kesslerloch; but this may arise from the peculiar character at that time of the land in the immediate neighbourhood. But when the stream had cut for itself a deeper channel than that near Thayngen, the water from the lake ran in this deeper channel, the water flowing from Radolphzell lake ceased to run, the valley became dry, and took its present form and size. From what has been said we may conclude that our cave was inhabited at a time when glaciers stretched far into Northern Switzerland. This period has been designated the 'Glacier Age.'

Whence they came—whither they went—these representatives of a people of long past ages—whether they followed the reindeer, which gradually withdrew into higher latitudes, or whether they were ejected by other conditions of nature unknown to us—who will venture to say? Only an unwearied investigation in this department will enable us to throw more light on the prehistoric age. Since then, mankind has pressed forward in every department of science and knowledge. This is known to us all. But the advance in its true magnitude is only recognised by us when we study such a portion of the history of civilisation as is presented to us in the Kesslerloch.

Das Alte fällt, es ändert sich die Zeit,
Und neues Leben blüht aus den Ruinen.

APPENDIX.

I. NOTE ON PLATE XV. FIGS. 98 AND 99.

In Mr. Merk's account, it will be seen that these two figures are very briefly mentioned, at the close of the report on the Thayngen fauna. This arose from their having come to light when the report was in the press; and after the full descriptions of the various objects had been printed off. Under these circumstances it will be well here to quote the description of them given by Professor Rütimeyer, in his late interesting work, ' Veränderungen der Thierwelt in der Schweiz seit Anwesenheit des Menschen' —a work which refers again and again to the Kesslerloch discoveries, and which is well worthy of careful study. He says (p. 98), ' The drawings of the bear and the fox are quite new; they were found, while this work was in the press, in the rubbish which came from the cave at Thayngen. They are represented with great truth in the woodcuts,' while the corre-

sponding figures, 98 and 99, of Mr. Merk do injustice to the original artist. The material is bone: and as far as I can tell by a careful comparison, they are on portions of ribs of the *Bison priscus*. The greater coarseness of the drawing arose from the very unfavourable nature of this hard and dry substance; and yet no one can help being struck by the point of Dr. Keller's remark when writing to me, that at all events the artist who ventured to draw the fox's head in front, and the bear sitting must have been a

---

¹ The cuts given here are copies of those in Professor Rütimeyer's book by the new ' Banks ' process.

different individual from the one who drew the reindeer and the horse, for the apparently defective amount of art in these two drawings evidently arises from the unfavourable nature of the material.'

So far Professor Rütimeyer: it will be seen that both he and also Dr. Keller accept these drawings as genuine. I am quite aware that it is a very serious matter even to doubt the verdict of such veterans of archæology; still it appears to me that, honestly, I ought here to give the singular history attached to them as it came to my knowledge when in Switzerland. During the progress of the excavation at the Kesslerloch the earth and stones were, as mentioned in the report, sifted in order to find the antiquities; the refuse was thrown aside, and in the process of the excavation became a considerable heap. After the excavation was finished, Mr. J. Messikomer, the well-known explorer of the lake-dwelling of Robenhausen, believed that a more careful sifting of the heap would bring to light many things which had escaped the earlier examination.

He therefore obtained permission from the owner of the soil to sift the mass again; and the result was an abundant harvest of flint flakes and bones of the reindeer, horse, Alpine hare, &c. Many of these are now in my possession. But unfortunately he employed for a stated period a labourer of the district, and when this period was expired, the man so employed says that he went to the heap on his own account, and discovered these two etchings of the bear and the fox! Those who are accustomed to the laws of evidence must decide, in the first place, whether the man committed any breach of honour or honesty, and in the next whether these etchings are genuine, or were made for the occasion. I do not venture to express a decided opinion, but I cannot help mentioning, on the one hand, that the style of the drawings is totally different from that of the other etchings; and on the other (and this appears rather a strong argument), the lines or furrows in the drawing of the bear have *within them* the ridge-like longitudinal projections, exactly like those on the plane surface, which have hitherto been attributed to weathering. It has been said that if a modern etching on bone is made when it is wet, this appearance would result; but having tried this plan on a piece of prehistoric bone from Bacon Hole, after it had been well soaked, and a copy of one of the doubtful specimens etched, with a Kesserloch flint, the lines engraved remained when dry perfectly sharp and clear, and there was no appearance *within* the furrows of the longitudinal fibrous appearance above referred to. The reader has now before him all the evidence which I can give on both sides, and he must judge for himself.

II. GLACIAL PERIOD.

It would be quite out of place here to give anything like a dissertation on this subject. It cannot, however, be wrong to state one or two facts which in a measure bear on the discoveries of the Kesslerloch. Although the Glacial period after the Tertiary is acknowledged by all thinking persons, yet every additional fact brought to light strengthens the theory—if in fact it can be now called by this name. The evidence of an Arctic or reindeer period in France is given so clearly in the valuable work called 'Reliquiæ Aquitanicæ,' that no one can gainsay it, an I it is needless to refer to a volume of such standing ; two other cases, one in Switzerland and the other in South Germany, may not be so well known.

In a hamlet called Schwerzenbach, which is reached by railway in about half an hour from Zürich, on the line to Wetzikon, there is a section of great interest, to which I was directed by Dr. Keller and guided by Mr. Jacob Messikomer. Here, on the banks of a small lake, and under a covering of about three feet of regular peat, there was a bed of about the same thickness of loam, and in this loam there were numerous leaves said to be of glacial age. The water unfortunately came in upon us when we dug towards the bottom of the loam, so that we could not reach the base nor ascertain what was below it, but Mr. Messikomer told me that it was what he called 'diluvium.' The loam bed is said to extend completely under the little lake. The leaves in the loam were in tolerable abundance, and apparently in fair preservation, and we brought away some masses of the clay, intending, if the specimens, after being soaked and dried, turned out well, to give a separate plate of them in the present volume. Unfortunately, however, when brought to England and soaked in water, no specimen worthy of being drawn could be secured. But the reader will have much better authority than a plate in the following extract from Professor Heer's Life of Arnold Escher Von der Linth, p. 261.

'A year ago a young Swedish geologist, Mr. Nathorst, discovered in the glacial clay of different parts of Sweden and Denmark a number of plants belonging to the extreme north, but which at that period must have lived in those districts, though at present they are only found in North Lapland and Spitzbergen. Mr. Nathorst last autumn came to Switzerland, to seek for this northern flora here, and he succeeded in finding it at Schwerzenbach, in the Canton of Zürich, in a bed of loam under a deposit of peat. With the assistance of Dr. Ferdinand Keller, who aided him by his rich fund of experience, masses of this loam under the peat were dug out and brought to Zurich ; and in it were found seven species of Arctic plants. They were, *Betula nana*, *Salix polaris*, *Salix retusa*, *Salix reticulata*, *Arctostaphylos uva ursi*, *Polygonum viviparum*, and *Dryas octopetala*. All these are species which belong to the far north, and, with the exception of the polar willow, live also in our Alps. The *Polygonum* and the *Dryas* are very plentiful at the foot of the glaciers, and the *Arctostaphylos* and *Salix retusa* and *S. reticulata* are not uncommon there, while the *Betula nana* prefers the peat moors. These plants evidently have descended from those which flourished at a time when the glaciers were retreating, and had formed tarns and small lakes like those of St. Gotthard. Like these the tarns were surrounded by an Alpine flora, the remains of which fell into the mud, and have been preserved to our own days. . . . In the 'Urwelt der Schweiz,' p. 534, &c., I have endeavoured to show that at the Glacial age the Alpine flora very probably occupied the plain land of Switzerland, so far as it was free from glaciers, and that the Alpine plants which we now see insulated on the mountain chains of Northern Switzerland may be the last remains of this ancient flora—the lost children of the Alps surrounded entirely by the inhabitants of the plain. This has now been confirmed by fresh discoveries, and doubtless further investigation will bring to light new facts, which will reveal to us clearly the plants and fauna of the Glacial age. But they have also in a most surprising manner confirmed another supposition, namely, that which considers our Alpine flora to have had its origin in the far north. Not only all the species which have as yet been found in the glacier mud have their home in the north, but amongst them is found the polar willow, which at the present day is only found in the Arctic zone (in North Lapland and Spitzbergen), but it is entirely wanting in the Alps. So that this species has come from the north into our districts, but here became extinct; while its companions have survived on the Alps up to the present day, and now are separated from their fellow species in the far north by a great extent of country.'

The young Swedish geologist above referred to, Mr. Nathorst afterwards visited Devonshire, in order to examine the Bovey deposits. He was delighted to find the *Betula nana* in the upper or white beds there. He is, I believe, the son of a Swedish professor, and is a very energetic naturalist : he mentioned that on his Spitzbergen expedition he had been three months without once having on dry clothes, and this had brought on total deafness. In this state he undertook and accomplished the journey to Switzerland and England to investigate these glacial plants.

More information as to the Bovey formations can be obtained in a paper on the lignite formation of Bovey by W. Pengelly, F.R.S., F.G.S., &c., and Professor Heer, printed in the 'Philosophical Transactions for 1863,' to which the reader is referred. At p. 26 (in the separate copies) will be found a note on the plants of the white clay or top beds, and these particulars are repeated, but somewhat more fully, in a note by Professor Heer in his Life of Escher, p. 279. He says, 'The white clay covering the lignite is of a totally different age from it. When the white clay was deposited *Betula nana* and some species of willow were living there, and their leaves have been preserved in this deposit. But *Betula nana* at the present day is not to be found in England, though it occurs on the Scotch mountains, and is very common in the Arctic zone. At the age of the glaciers it was also living in the south of Sweden, in Denmark, and on the plain-land of Switzerland, where its remains have been found together with those of other Arctic plants in the deposits of this period. So that there can be no doubt that at that time Devonshire had a colder climate than at present, and that when the temperature changed it was forced back to the Scotch mountains.

The young Swedish naturalist and his discoveries have led us away from Mid-Europe, and we must now return for a few moments to a locality in Upper Swabia, north of the Lake of Constance, as discoveries of the reindeer age were made there some years ago which threw considerable light on the contents of the Kesslerloch. They were fully described by Professor Fraas in the 'Württembergische naturwissenschaftliche Jahreshefte' for 1867, and a somewhat shorter notice of them, also by Professor Fraas, was given translated into English in the 'Geological Magazine' for 1866, p. 546. The reader is therefore referred to these two accounts: but it may perhaps be well to mention a very few facts respecting the place. Schussenried is a small hamlet to the south of the rise forming the watershed in South Germany to the north and the south: the streams on the south running into the Rhine, those on the north into the Danube. Under a considerable depth of peat near this village there was a bed, of four or five feet thick, of what Professor Fraas calls tufa sand, or loose tufa, beneath which was the relic-bed, also four or five feet thick, containing a very large quantity of reindeer-horns, and of tools and implements made from them, also of flint and stone implements. One of the reindeer-horns has upon its palm certain indefinite etchings, but by no means distinct enough to be called a drawing. This was shown to me by Professor Fraas when at Stuttgard on my return from Switzerland. It is somewhat singular that almost everything, whether needle or scraper or any other tool, found here was either broken or unfinished,[1] so that it appears as if this accumulation had been merely a rubbish-heap of rejected implements. Together with these there were several species of *Hypnum*, which, according to Professor Schimper of Strassburg, who has paid much attention to mosses, consist entirely of northern or high Alpine forms. It will be seen that the objects found at Schussenried bear a great similarity to those from the Kesslerloch.

Sufficient has now been said to show that an Arctic climate at one time extended not merely over a great part of France, but also over Middle Europe. It appears to me that only those who are resolved not to be convinced can doubt the evidence of Aquitaine, of Thayngen, and of Schussenried.

[1] I cannot resist copying here a few lines from an interesting volume just published by Miss Isabella L. Bird, called *Six Months amongst the Palm Groves, Coral Reefs, and Volcanoes of the Sandwich Islands*. She is giving an account of her ascent of the volcanic mountain in Hawaii, Mauna Kea, 13,953 feet above the sea. After leaving the summit, she came to 'a cave, a lava-bubble, in which the natives used to live when they came up here to quarry a very hard adjacent phonolite for their axes and other tools.' She says, 'I was glad to make it a refuge from the piercing wind. *Hundreds of unfinished axes lie round the cave entrance, and there is quite a large mound of unfinished chips.* This is a very interesting spot to Hawaiian antiquaries. They argue from the amount of the chippings that this mass of phonolite was quarried for ages by countless generations of men, and that the mountain top must have been upheaved, and the island inhabited in a very remote past. The stones have not been worked since Capt. Cook's day.'—pp. 351, 352.

May not this account throw some light on what are called by antiquaries 'rubbish heaps' of implements?

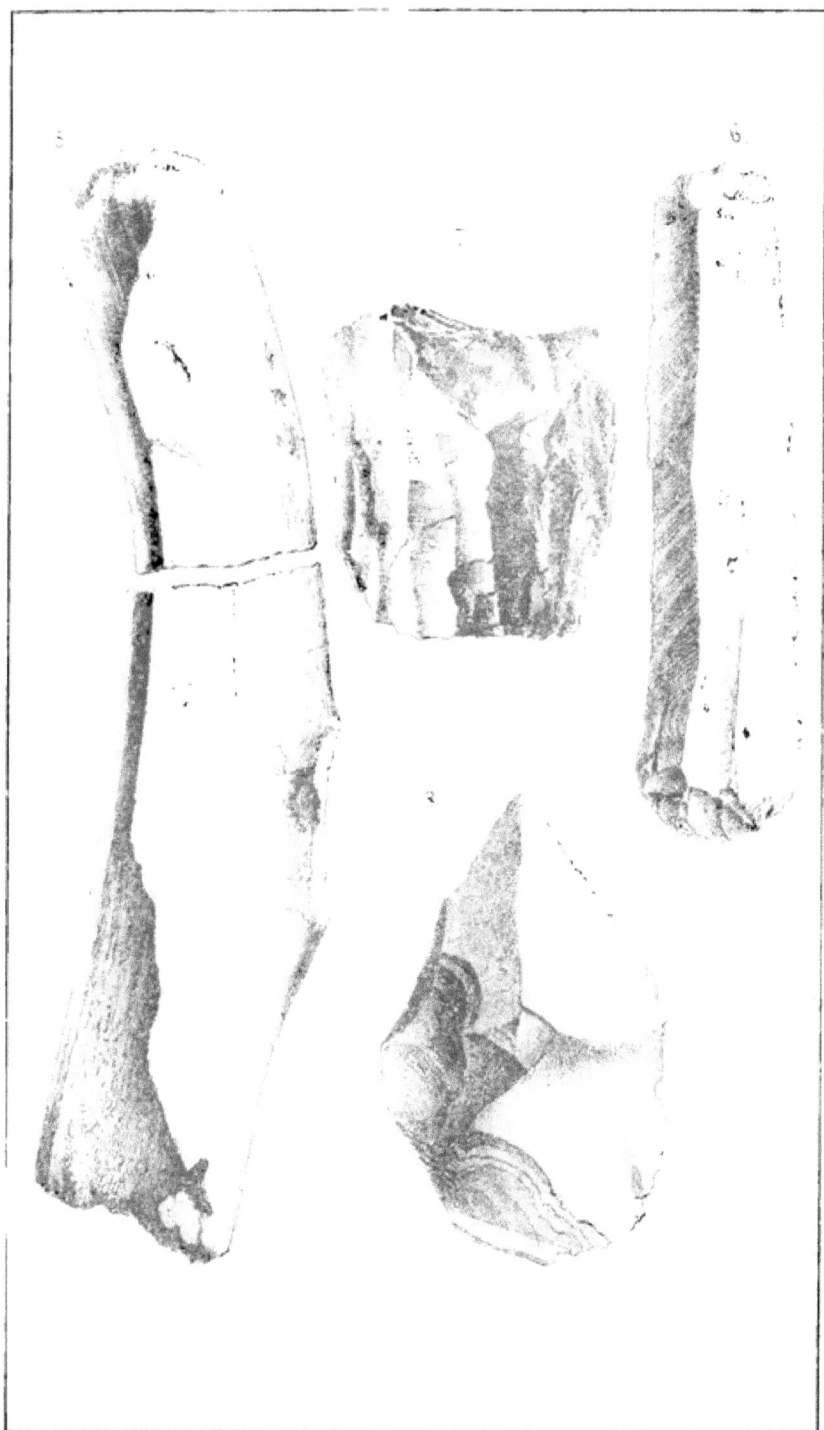

Pl II.

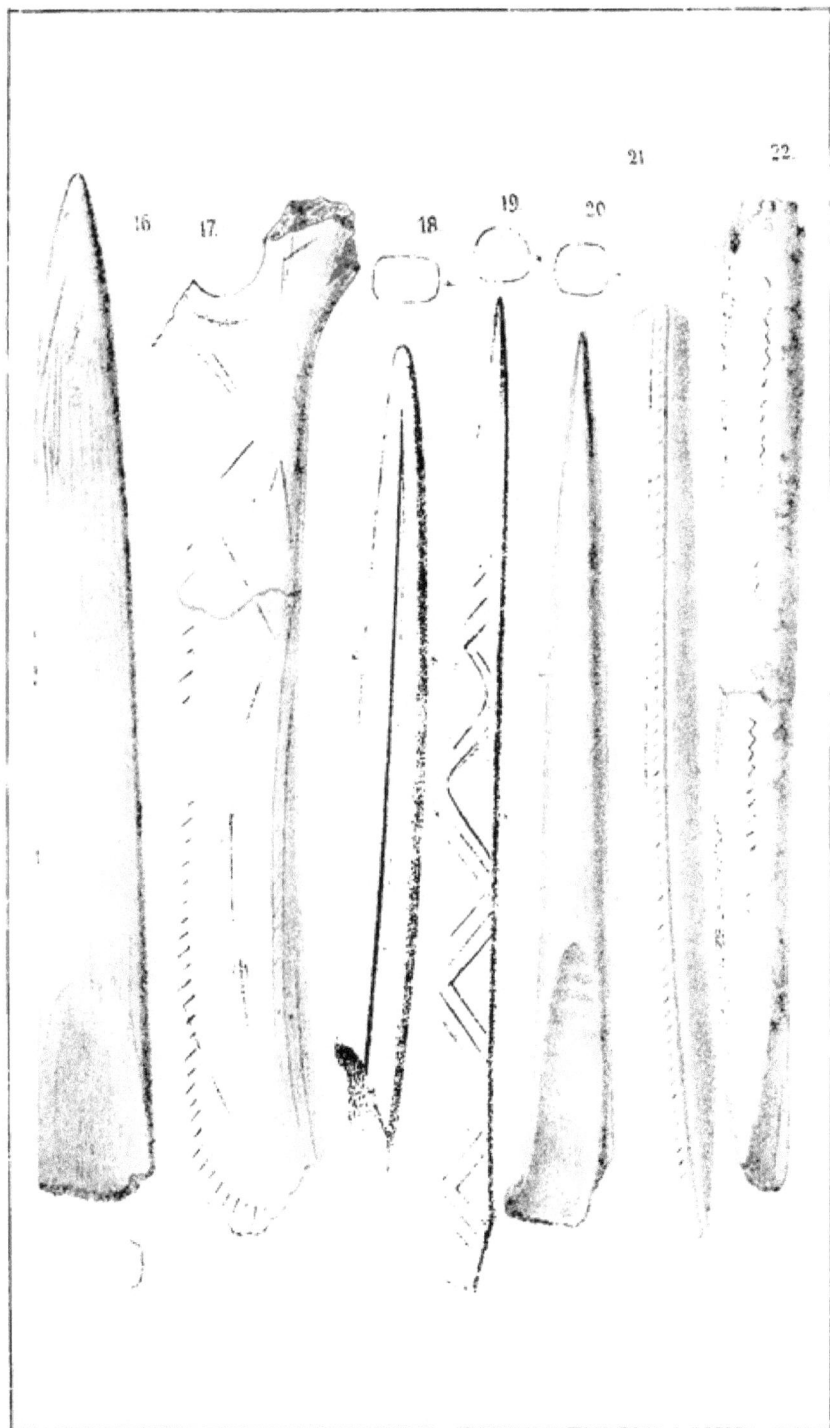

16. 17. 18. 19. 20. 21. 22.

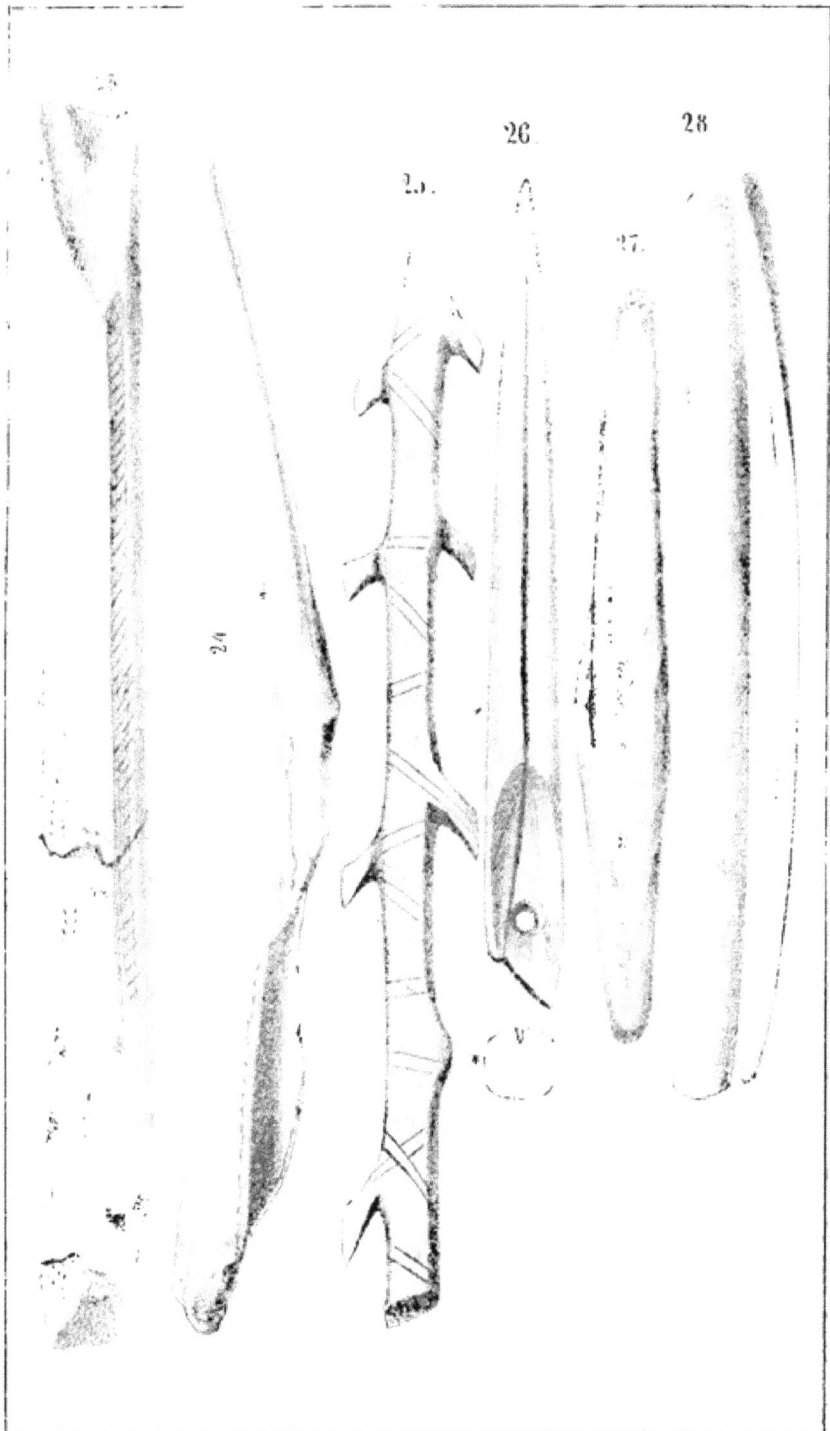

Pl. VII.

29. 30. 31. 32. 33. 34. 35

39.

37.

38.

40.

36

41

42

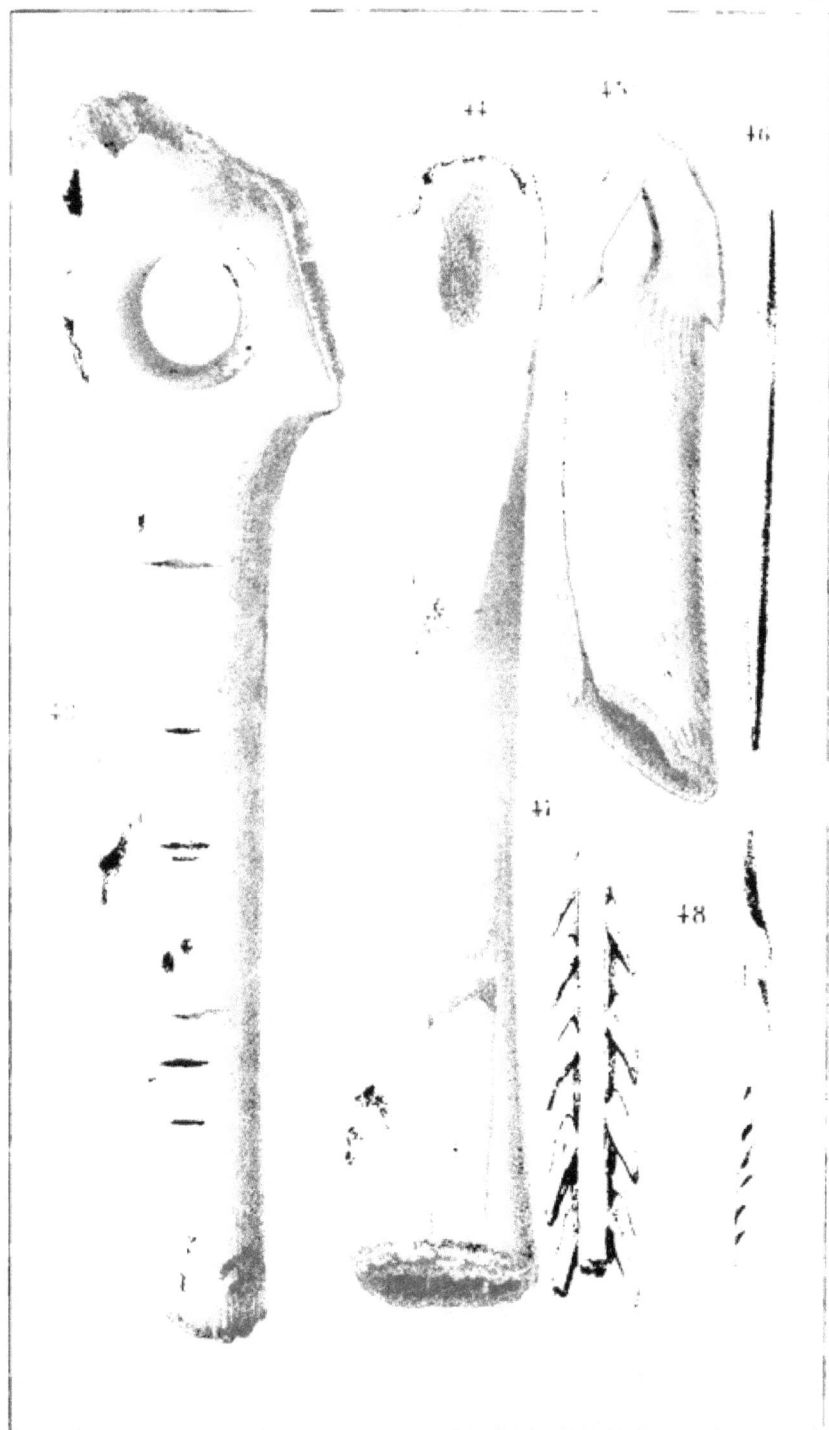

44　45　46

43

47

48

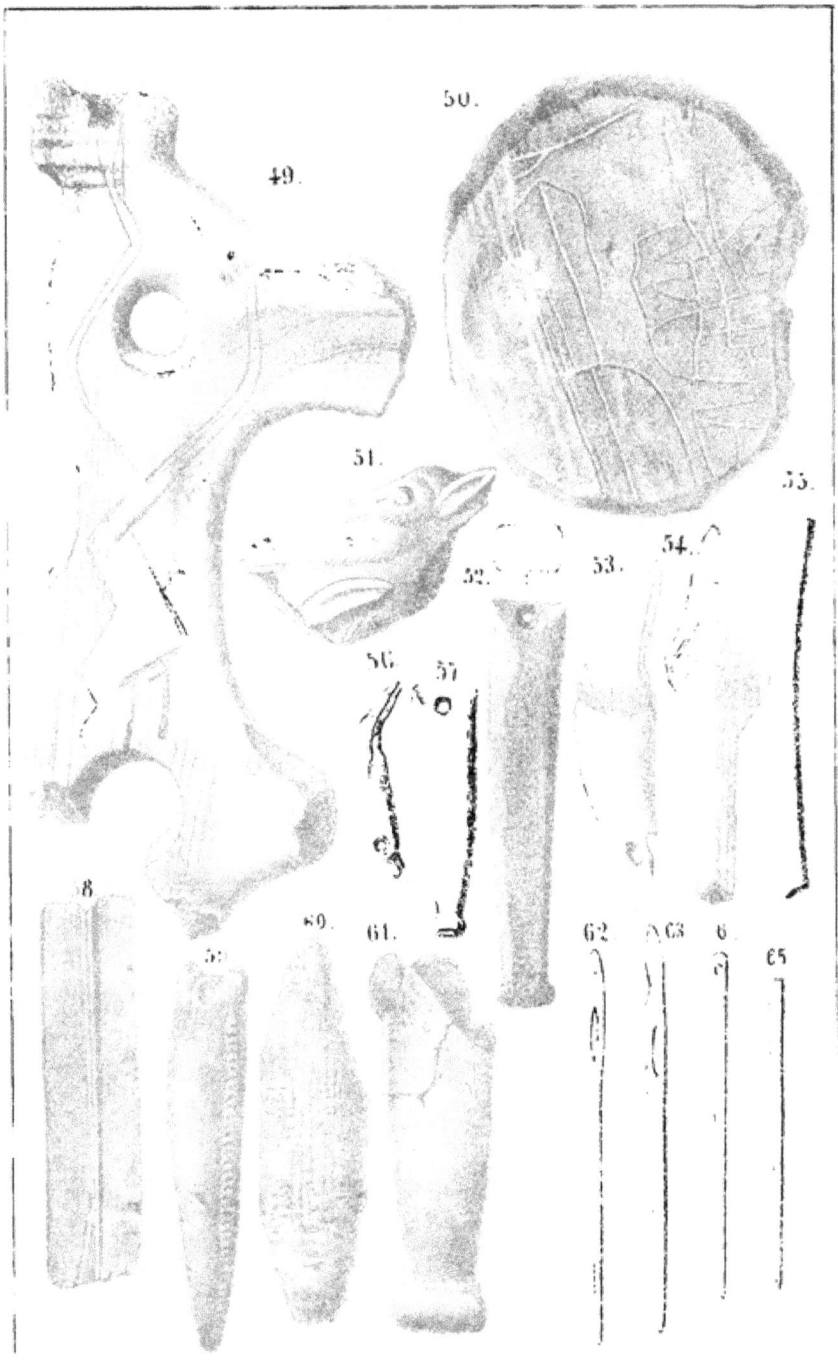

Pl IX

49.

50.

51.

55.

52.

53.

54.

56.

57.

58.

59.

60.

61.

62.

63.

6.

65.

Pl X

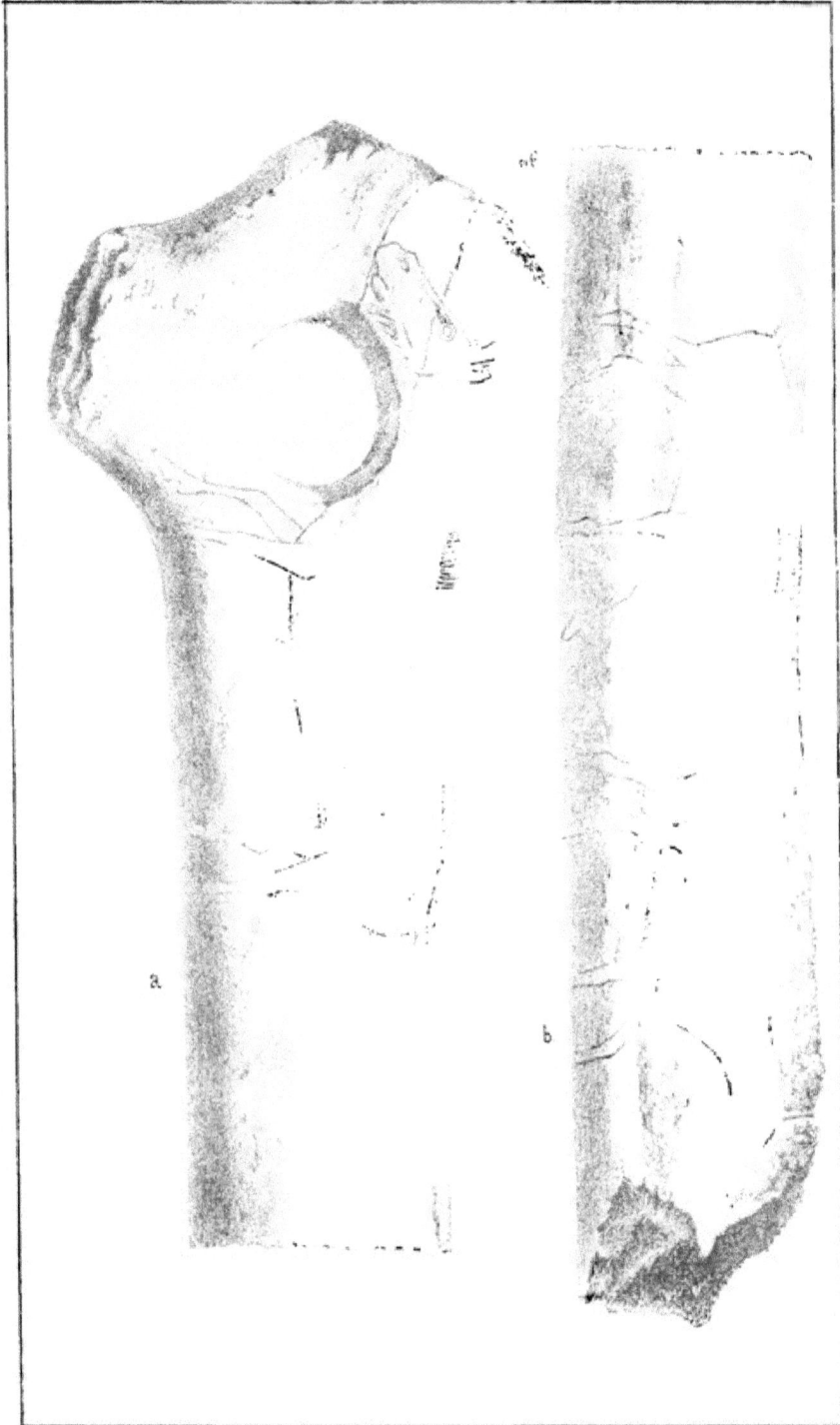

a.   b.

68

a

69

b

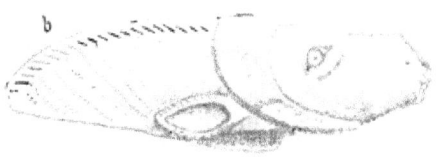

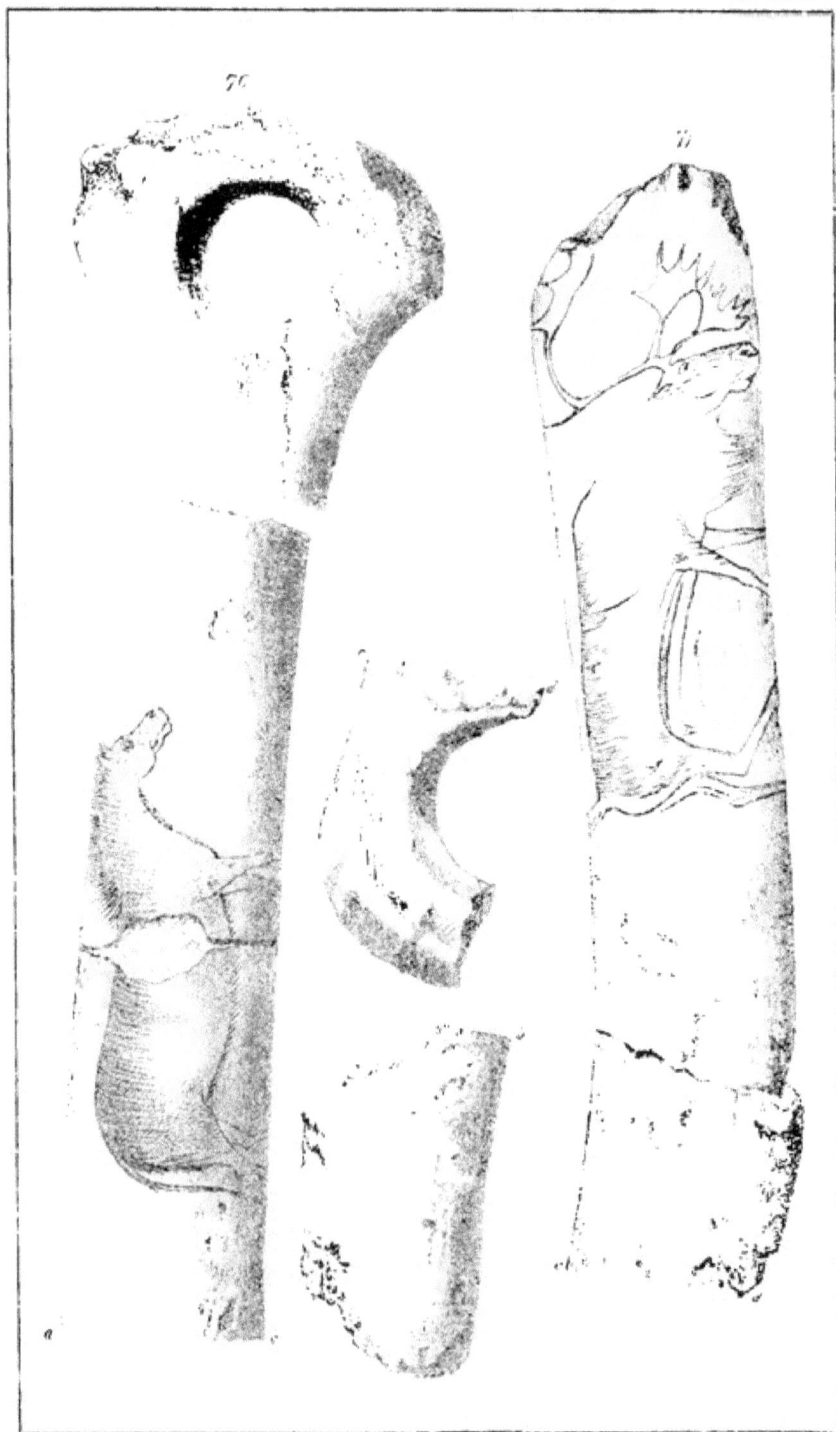

Pl. XII.

Pl. XII

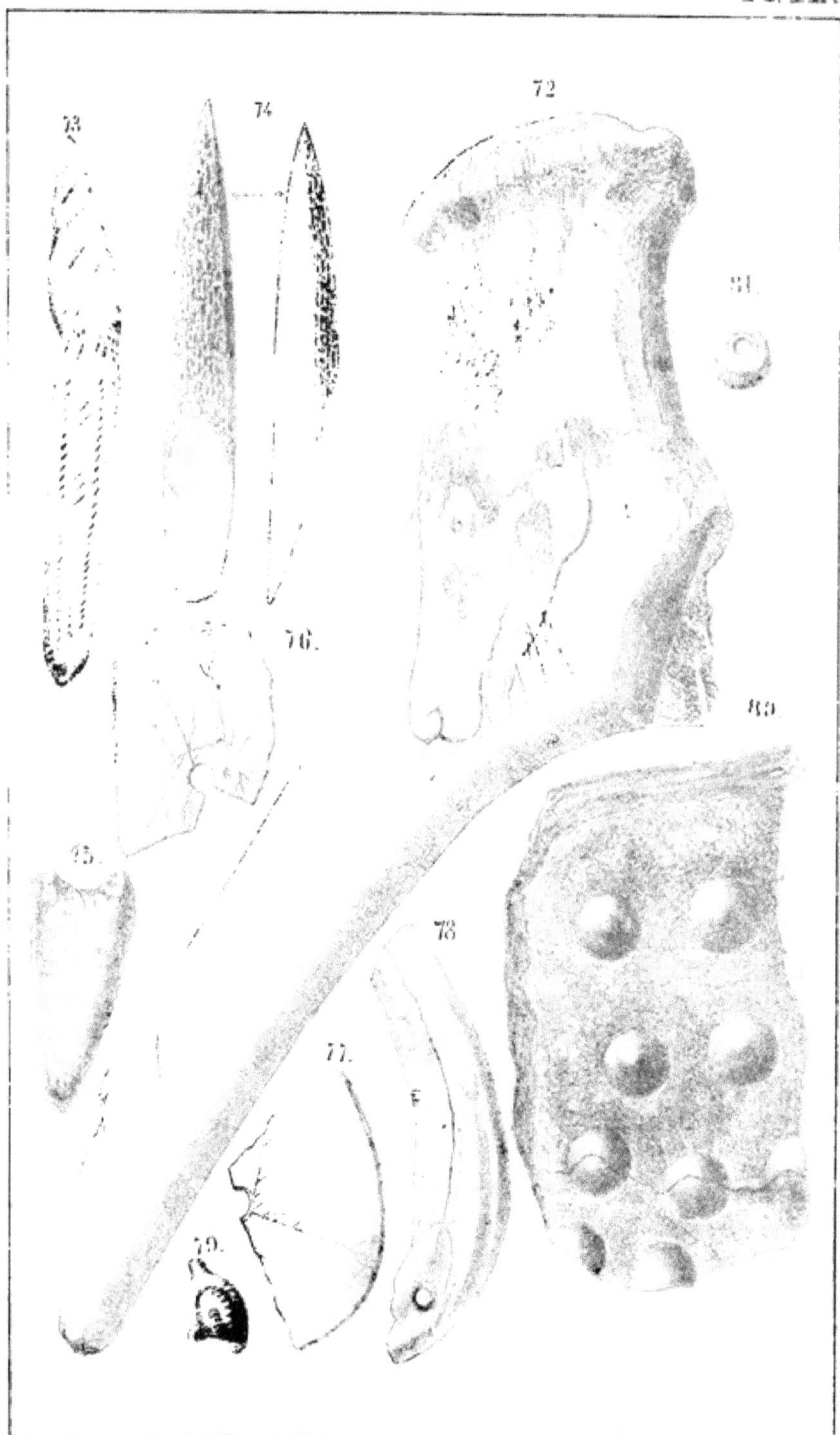

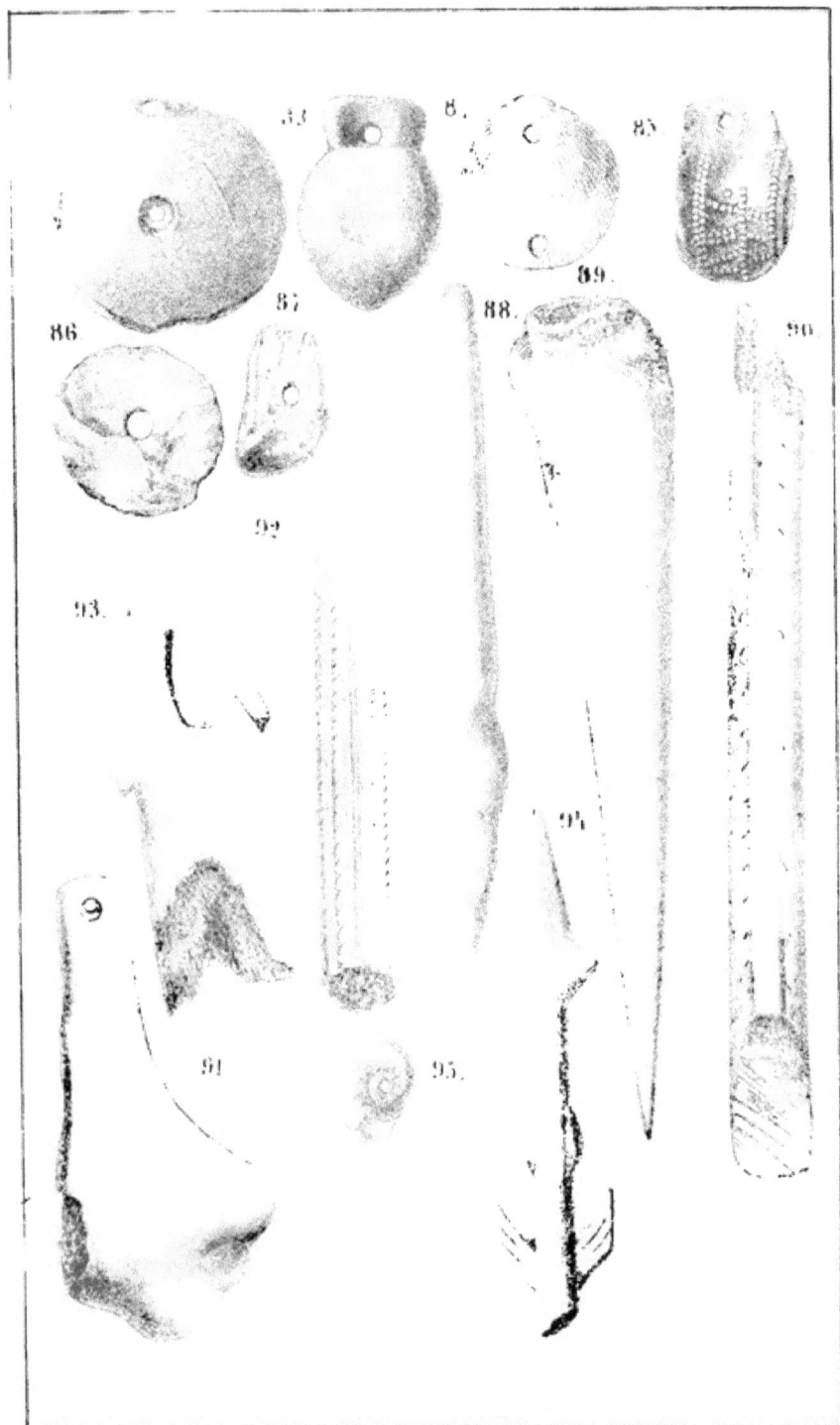

Pl. XIV.

96.

97.

98.

99.

EXPLANATION OF THE PLATES.

head, ornamented, p. 35.—34. Bone arrow-head, p. 34.—35. Bone harpoon, p. 38.
—36. Broken spear- or arrow-head, p. 35.—37. Perforated ammonite, p. 45.—
38, 39, 40. Ornaments of stone, p. 45.—41. Bâton (?) with two perforations,
p. 46.—42. Ornamented poignard, p. 43.

Plate VIII.

Fig. 43. Perforated Bâton (?) p. 46.—44. Implement (for extracting roots
from the ground?), p. 43.—45. Ornamented bone implement, p. 43.—46. Bone
implement sharpened at both ends, p. 42.—47. Bone harpoon, p. 39.—48. Bone
harpoon, p. 38.

Plate IX.

Fig. 49. Bone with two perforations, p. 46.—50. Indefinite drawing on
brown coal, p. 48.—51. Carved head of horse, p. 52.—52. Earring of slate coal,
p. 45.—53. Perforated tooth, p. 43.—54. Point of a harpoon, p. 38.—55. Unknown implement, p. 43.—56. Perforated horse's tooth, p. 43.—57. Bone earring, p. 44.—58. Piece of brown coal with an incision lengthwise, p. 44.—59,
60, 61. Earrings of brown coal, p. 45.—62, 63, 64, 65. Bone needles, p. 41.

Plate X.

Fig. 66a and b. Drawings of a horse and two reindeer on the 'beam' of a
reindeer-horn, p. 48.

Plate XI.

Fig. 67. Drawing, probably of a pig, p. 46.—68. Drawing of a horse (see fig.
70), p. 50.—69a and b. Carved head of musk-sheep, p. 51.

Plate XII.

Fig. 70. Drawing of a horse on the branch of a reindeer-horn, p. 50.—71.
Drawing of a reindeer on the branch of a reindeer-horn, p. 49.

Plate XIII.

Fig. 72. Reindeer-head drawn on a dagger-like bone, p. 47.—73. Unknown
bone implement, p. 42.—74. Small bone spear-head, p. 34.—75. Earring of
brown coal, p. 45.—76, 77. Ornamented plate of bone, p. 44.—78. Perforated
horse's tooth, p. 43.—79. Earring of brown coal, p. 45.—80. Piece of pottery
found in the uppermost bed, p. 52.—81. Earring of brown coal, p. 45.

Plate XIV.

Fig. 82, 83. Earrings of brown coal, p. 44.—84. Perforated pectunculus,
p. 45.—85. Earring of brown coal, p. 44.—86. Earring (?) of slate coal, p. 44.—
87. Ground tooth, p. 43.—88. Unknown bone implement, p. 42.—89. Awl,
p. 40.—90. Ornamented spear-head, p 37.—91. Perforated humerus, p 43.—92.
Ornamented bone implement, p. 43.—93. Unknown bone implement, p. 43.—
94. Bone harpoon, p. 39.—95. Perforated ammonite, p. 45.

Plate XV.

Fig. 96, 97. Drawing of a horse's head on brown coal, p. 47.—98. Drawing
of a bear, p. 61.—99. Drawing of a fox, p. 61.

Spottiswoode & Co., Printers, New-street Square, London.